Graphene-Based Materials
Science and Technology

T0225412

Graphene-Based Materials
Science and Technology

Subbiah Alwarappan
Ashok Kumar

CRC Press
Taylor & Francis Group
Boca Raton London New York

CRC Press is an imprint of the
Taylor & Francis Group, an **informa** business

CRC Press
Taylor & Francis Group
6000 Broken Sound Parkway NW, Suite 300
Boca Raton, FL 33487-2742

First issued in paperback 2017

Version Date: 20130822

ISBN 13: 978-1-4398-8427-0 (hbk)
ISBN 13: 978-1-138-07450-7 (pbk)

Visit the Taylor & Francis Web site at
http://www.taylorandfrancis.com

and the CRC Press Web site at
http://www.crcpress.com

Contents

Preface

An atom-thin form of carbon called graphene has been continuously studied since its discovery due to its versatile properties, such as exceptional electronics, half-integer quantum hall effect, ballistic electron transport, optoelectronic properties, and high crystal quality. Further, graphene offers truly unique opportunities because, unlike most semiconductor systems, its two-dimensional (2D) electronic states are not buried deep under the surface, and it can be easily accessed directly by tunneling or by other local probes. Until now, graphene has been considered the strongest and thinnest material known. Tremendous advancements have been made for the application of graphene in nanoelectronics. Moreover, graphene has been used in biological systems for the detection of DNA, RNA, proteins, and nucleic acids. In addition, graphene-based nanodevices are available for the detection of bacteria and pathogens. Presently, graphene holds the key to everything from small computers to high-storage batteries and capacitors. Graphene's properties are attractive to physicists, materials scientists, and electrical engineers for several reasons, not least of which is that it might be possible to build circuits that are smaller and faster than what can be built in silicon. Due to all the properties exhibited by graphene, A. K. Geim and K. S. Novoselov, the inventors of graphene, were awarded a Nobel Prize in 2010. In this book, we analyze the recent advancements in graphene research, such as synthesis, properties, and important applications in several fields.

Chapter 1 gives a brief introduction regarding the history of graphene and its important properties. Chapter 2 discusses the different methods of graphene synthesis available in the literature. Chapter 3 gives a brief overview of a few important characterization techniques that distinguish graphene from its allotropes. The application of graphene in gas sensors is presented in detail in Chapter 4. In Chapter 5, the application of graphene in biosensors and energy storage is discussed in detail. Chapter 6 presents the important photonic device and optoelectronic device applications of graphene-based materials.

We would like to thank Professor Shyam Mohapatra, distinguished professor and director, Nanomedicine Research Center, College of Medicine, University of South Florida, Tampa, for his valuable suggestions and comments during the preparation of this book.

Dr. Alwarappan would also like to thank Chen-Zhong Li, Kaufmann professor, Department of Biomedical Engineering, Florida International University, Miami, with whom he started his graphene-based electrochemical biosensor research. Since 2009, Dr. Alwarappan has been supported by Dr. Li in his graphene-related work.

We would like to extend thanks to the technical staff at the Nanotechnology Research and Education Center and Nanomedicine Research Center, University of South Florida, Tampa, for their support in our graphene research for the last two years.

We extend thanks to the publishers who granted us permission to reproduce the artwork and experimental figures published in their journals.

We are grateful to our sponsors for funding our research.

Finally, Dr. Alwarappan would like to thank his wife and family for their support and encouragement throughout the preparation of this book.

Dr. Subbiah Alwarappan
Nanotechnology Research and Education Center
College of Engineering
University of South Florida, Tampa, Florida

Prof. Ashok Kumar
Nanotechnology Research and Education Center
College of Engineering
University of South Florida, Tampa, Florida

About the Authors

Dr. Subbiah Alwarappan received his BSc (chemistry, 1999) from Alagappa Government Arts College, Karaikudi, Tamilnadu, India, and MSc (chemistry, 2001) from Presidency College, Chennai, Tamilnadu, India. Later, he was awarded an International Postgraduate Research Scholarship fellowship to pursue his PhD (in electroanalytical chemistry, 2006) at Macquarie University, Sydney, Australia. During his PhD studies, he worked toward the design of miniaturized pyro-lyzed carbon electrodes for the *in vivo* detection of important neurotransmitters. Later, he worked as a postdoctoral researcher at the University of Iowa, Iowa City, for one year. In November 2007, he moved to Florida International University, Miami, and spent two years there as a postdoctoral research associate (November 2007 to November 2009). He accepted a position at the Nanomaterials Research and Education Center, University of South Florida, Tampa, and worked there for a year as a senior postdoctoral researcher (November 2009 to December 2010). He was then a joint research faculty member at the College of Medicine and Nanotechnology Research and Education Center at the University of South Florida, Tampa (January 2011 to January 2013). During this period, his research interests included synthesis and characterization of novel carbon-based materials such as graphene, carbon nanotubes for high-performance biosensing, immunosensing, environmental toxin detection, and modeling of various pro-cesses occurring at an electrode's interface. Dr. Alwarappan has published more than 26 peer-reviewed research articles in the electroanalytical research area, especially on graphene-based electrochemical sensors. He also authored three book chapters and has delivered more than 20 presentations at conferences, symposia, and invited talks. His articles have been cited more than 500 times. He has been the invited reviewer for more than 30 peer-reviewed journals published by the Royal Society of Chemistry (RSC), American Chemical Society (ACS), and Elsevier. He is also serving as a reviewer for several prestigious funding agencies.

Dr. Ashok Kumar is a professor in the Department of Mechanical Engineering at the University of South Florida, Tampa. His research is focused toward the development of nanotechnology-based novel materials for multifunctional applications. His other interests include K–12 educational outreach, gender and science education, and nanotechnology industrial outreach. He has published 2 textbooks; edited 7 books of proceedings; and 12 book chapters, including 200 peer-reviewed articles. His excellence as a researcher has been recognized with a number of honors, including the ASM Fellow (2007), AAAS Fellow (2010), ASM-IIM Visiting Lecture Award (2007), Theodore and Venette Askounes Ashford Distinguished Scholar Award (2006), USF Outstanding Faculty Research Achievement Award (2004), USF President Faculty Excellence Award (2003), NSF (National Science Foundation) Faculty Early Career Development Award (2000), National Research Council Twining Fellowship Award (1997), and NSF and DOE (Department of Energy) EPSCoR Young Investigator Awards (1996–1997). He also received the Professor Honorario award from the Universidad del Norte, Barranquilla, Colombia (2009) and an outstanding faculty award in 2013 from the University of South Florida.

Chapter 1

Graphene: An Introduction

1.1 Graphene: History and Background

The possibility of graphene's existence or that of a two-dimensional (2D) allotrope of carbon has been theoretically studied for 60 years. Often, the term *graphene* was used to describe the properties of carbon allotropes [1–3]. However, after four decades it has been realized that graphene also provides an excellent condensed matter analogue of (2 + 1)-dimensional quantum electrodynamics [4–7], thereby exposing graphene to a thriving theoretical "toy" model [7]. Graphene was expected to be unstable due to the formation of curved structures such as soot, fullerenes, and nanotubes. Further, graphene was believed not to exist in its free state. Unexpectedly, in 2004, the prediction of graphene's existence became true when freestanding graphene was discovered by Geim and Novoselov [8,9]. Moreover, the follow-up experiments demonstrated that its charge carriers were indeed massless Dirac fermions [10,11]. As a result of this phenomenon, graphene is indeed the material of choice for numerous researchers. Geim and Novoselov shared the 2010 Nobel Prize in Physics for the discovery of graphene [12–14].

Graphene is considered the flat single layer of carbon atoms that are tightly packed into a honeycomb-like crystalline lattice in a 2D fashion [8–11]. Further, graphene is often known as the "mother" or the basic building unit of all other carbon allotropes. For instance, graphene can be wrapped up to a zero-dimensional (0D) fullerene, rolled to resemble one-dimensional (1D) carbon nanotubes, or stacked to a three-dimensional (3D) graphite (see Figure 1.1) [8–11]. To understand 2D graphene in detail, we briefly describe the 2D crystals [7]. A single atomic plane of graphene is

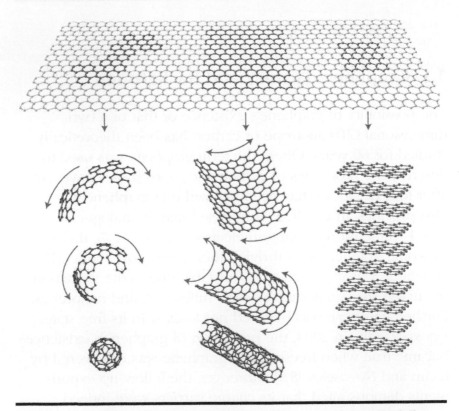

Figure 1.1 **Scheme showing graphene can be wrapped to 0D fullerenes, wrapped to form 1D carbon nanotubes (CNTs), or stacked to form 3D graphite. (Reproduced with permission from A.K. Geim, K.S. Novoselov, *Nat. Mater.* 6, 183–191, 2007.)**

considered a 2D crystal, whereas graphene with more than 100 layers is usually considered a thin film of a 3D material. However, there always exists a question: How many layers will make this a 3D structure? In graphene, the electronic structure has been found to evolve rapidly with the number of layers, and it will approach the 3D limit of graphite at exactly 10 layers [15].

Single-layer graphene (SLG) and bilayer graphene have simple electronic spectra; both are zero-gap semiconductors (also referred as zero-overlap semimetals) with one type of electron and one type of hole. On the other hand, the spectra of graphene containing three or more layers become too complicated. Further, with these three or more layers, several charge carriers are noticed. In addition, the conduction and valence bands begin to overlap [8,15,16]. Considering all these features, it is easy to distinguish SLG and double- and few- ($3 \leq 10$) layer graphene as three different types of 2D crystals ("graphenes").

In graphite, the screening length is only about 5 Å (less than two layers thick). So, researchers should differentiate the surface and the bulk even for films as thin as five layers [15,16]. Concentrated graphene has been isolated by chemical exfoliation of intercalating bulk graphite and causes the separation of graphene planes by layers of intervening atoms or molecules and results in the formation of new 3D materials [17]. However, in certain cases, large molecules could be inserted between atomic planes, providing greater separation, such that the resulting compounds could be considered as isolated graphene layers embedded in a 3D matrix. Furthermore, one can often get rid of intercalating molecules in a chemical reaction to obtain a sludge consisting of restacked and scrolled graphene sheets [18,19]. Because of its uncontrollable character, graphitic sludge has so far attracted only limited interest. Graphene has wide potential applications due to its excellent mechanical, electrical, thermal, and optical properties and its large surface-to-weight ratio (e.g., 1 g of graphene can cover several football fields) [12,20]. The remarkable properties of graphene reported in the literature are as follows: Young's modulus, about 1,100 GPa [21]; fracture strength, 125 GPa [21]; thermal conductivity, about 5,000 W m^{-1} K^{-1} [22]; mobility of charge carriers, 200,000 cm^2 V^{-1} s^{-1} [23]; and specific surface area calculated value, 2,630 m^2 g^{-1} [24]. It has fascinating transport phenomena, such as the quantum Hall effect (QHE) [25]. The exceptional thermal, optical, and electrical property of graphene is the result of its extended π-π conjugation [26]. It is also worth mentioning that graphene possesses a number of surface-active functional moieties, such as carboxylic, ketonic, quinonic, and C=C. Of these, the carboxylic and ketonic groups are reactive and can easily bind covalently with several biomolecules, thereby influencing the possibility of functionalizing graphene with biomolecules for various biosensing applications [27–30]. In addition, reports confirmed that graphene and chemically modified graphene (CMG) are promising candidates for energy storage materials [24], paper-like materials [31,32], polymer composites [33,34], liquid crystal devices [35], and mechanical resonators [36–38].

The term *graphene* was recommended by the IUPAC (International Union of Pure and Applied Chemistry) commission to replace the existing *graphite layers* term that was irrelevant in the single-carbon-layer structure research because a 3D stacking structure is identified as *graphite* [39]. As a result, these days *graphene* refers to a 2D monolayer of carbon atoms and is considered as the basic building block of graphitic materials (i.e., fullerene, carbon nanotube, and graphite) [39].

1.2 Graphene Properties

Since 2006, several fascinating properties have been evident from pristine graphene. The exciting properties of graphene include high charge (electrons and holes) mobility (230,000 cm^2/Vs) with the absorption of visible light up to 2.3%, thermal conductivity (3000 W/mK), high strength (130 GPa), and high theoretical specific surface area (2600 m²/g) [37–39]. Further, graphene exhibits a half-integer QHE even at ambient temperature (minimum Hall conductivity 4 e²/h, even at zero carrier concentration), which has filled the research community with great enthusiasm [39]. Here, we focus on and discuss the properties that constitute the basic foundation for graphene's wider scope of applications.

1.2.1 Electrical Transport Property

Pristine graphene is a zero-gap semiconductor [39]. The sp² hybridized carbon atoms are arranged in hexagonal fashion in a 2D layer. A single hexagonal ring contains three strong in-plane sigma bonds perpendicular to the planes [39]. Different graphene layers are bonded by weak pz interaction, whereas strong in-plane bonds keep hexagonal structure stable and facilitate the delamination of 3D structure (graphite) into individual graphene sheets by the application of a mere mechanical stress [39]. As discussed previously, single-layer, defect-free graphene can be obtained by the Scotch tape method through micromechanical cleavage [39]. This route provides a 2D platform that is a basis for studying several fundamental properties of this crystal.

An interesting fact about graphene is the anomalous behavior of its charge carriers, which behave as massless relativistic particles (Dirac fermions) [39]. In general, the behavior of Dirac fermions is abnormal in comparison with electrons under a magnetic field. For example, the anomalous integer quantum Hall effect (IQHE) was noticed even at room temperature [39–42]. The charge carriers in graphene have a distinctive nature that mimics relativistic particles, which are considered electrons without their rest mass. These particles can be better described with a (2 + 1)-dimensional Dirac equation [7,39]. The band structure of SLG is composed of two bands intersecting at two in-equivalent points K and K⁰ (K and K⁰ are the Dirac points where valence and conduction

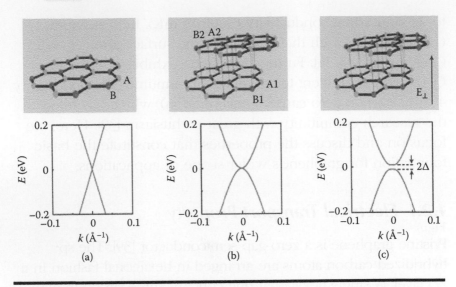

Figure 1.2 (See color insert.) Band gap in graphene. Schematic diagrams of the lattice structure of (a) monolayer and (b) bilayer graphene. The green and red lattice sites indicate the A (A1/A2) and B (B1/B2) atoms of monolayer (bilayer) graphene, respectively. The diagrams represent the calculated energy dispersion relations in the low-energy regime and show that monolayer and bilayer graphene are zero-gap semiconductors. (c) When an electric field E is applied perpendicular to the bilayer, a band gap is opened in bilayer graphene, whose size (2Δ) is tunable by the electric field. (Reproduced with permission from J.B. Oostinga, H.B. Heersche, X. Liu, A.F. Morpurgo, L.M.K. Vandersypen, *Nat. Mater.* 7, 151–157, 2008.)

bands degenerate, thereby making graphene a zero-bandgap semiconductor) (Figure 1.2). The electronic conductivity of graphene is directly related to the quality of the graphene. For example, the higher the quality of graphene is (i.e., low defect density of its crystal lattice), the greater will be its conductivity. In general, defects act as scattering sites and block the charge transport by limiting the mean free path of the electron. Based on evidence, pristine graphene is free from defects, and its conductivity is affected by a number of extrinsic sources [39].

The major factors that affect the conductivity of graphene include surface charge traps, interfacial phonons, and substrate ripples [39,43–46]. According to Kim and coworkers [37],

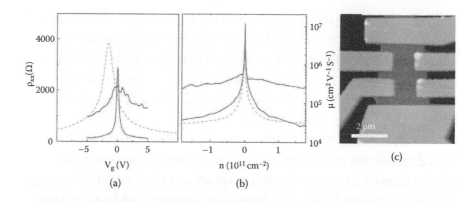

Figure 1.3 (See color insert.) (a) Four-probe resistivity ρ_{xx} as a function of gate voltage V_g before (blue) and after (red) current annealing; data from a traditional high-mobility device on the substrate (gray dotted line) is shown for comparison. The gate voltage is limited to the ±5-V range to avoid mechanical collapse. (b) Mobility, $\mu = 1/en\,\rho_{xx}$ as a function of carrier density n for the same devices. (c) Atomic force microscopy (AFM) image of the suspended graphene before the measurements. (Reproduced with permission from K.I. Bolotin, K.J. Sikes, Z. Jiang, M. Klima, G. Fudenberg, J. Hone, et al., *Solid State Commun.* 146, 351–355, 2008.)

the mobility exceeds 200,000 cm²/Vs at a carrier density of 2×10^{11} cm⁻² for a mechanically exfoliated suspended layer of graphene above a Si/SiO₂ gate electrode (Figure 1.3). Further, substrate-induced scattering was minimized by employing suspended SLG, and the mobility has been improved to 230,000 cm²/Vs by inducing electric current. Another important characteristic of SLG is that charge carriers can be tuned between electrons and holes by applying a required gate voltage; this is called ambipolar electric behavior [7,8,39]. During positive gate bias, the Fermi level rises above the Dirac point and promotes electrons populating into the conduction band, whereas under the negative gate bias condition, the Fermi level drops below the Dirac point and promotes holes in the valence band in concentrations of $n = \alpha Vg$ (where α is a coefficient depending on the SiO₂ layer used as a dielectric in the field effect devices, and Vg is the gate voltage).

Researchers consider graphene key material for the future generation of electronic devices. Graphene suffers from zero-energy band gap even at the charge neutrality point, which is one of the disadvantages of graphene if employed as an electronic material [39]. The zero band gap restricts its use in logic applications and requires frequent on/off switching. However, the band structure of graphene can be altered by lateral quantum confinement by constraining the graphene within nanoribbons or graphene quantum dots and by biasing bilayer graphene [39,47–53]. Band gap opening in both zigzag and armchair nanoribbons was noticed. Further, this has been proven both experimentally and theoretically, and it varies with the width of the ribbons and disorder in the edges [39,54]. Doping and edge functionalization alter the band gap in nanoribbons [39,55]. It is important to note that several primary works on graphene has been related to field-effect transistors (FETs). A schematic of the graphene-based FET is shown in Figure 1.4. It consists of the gate, a graphene channel connecting source, drain electrodes, and a dielectric barrier layer (SiO$_2$) separating the gate from the channel. In several studies, a 300-nm SiO$_2$ layer was employed. Graphene acted

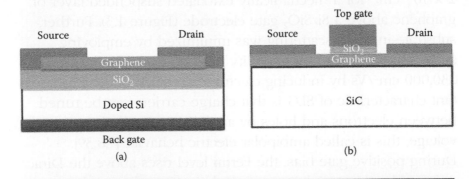

Figure 1.4 (See color insert.) (a) Schematic diagram of back-gated and (b) top-gated graphene field effect devices. The perpendicular electric field is controllable by applied back-gate voltage V_g and the top-gate voltage V_{top}. (Reproduced with permission from V. Singh, D. Joung, L. Zhai, S. Das, S. I. Khondaker, S. Seal, *Prog. Mater. Sci.* 56, 1178–1271, 2011.)

as a dielectric layer, and silicon acted as a back gate. Due to the large parasitic capacitance of the 300-nm SiO_2 layer, it is difficult to integrate with other electronic components. As a result, top-gated graphene-based devices have been developed [39,56]. Top-gated graphene-based metal-oxide semiconductor FETs (MOSFETs) have been designed with exfoliated graphene [39,56–59], chemical vapor deposition (CVD) grown graphene on nickel and copper substrate, and epitaxial graphene. In addition, materials such as Al_2O_3, SiO_2, and HFO_2 dielectric materials have been used for the top gate [39,53,60–63].

Large-area epitaxial graphene has attracted wider recognition due to its scalability for electronics [39]. Few-layer graphene films grown on SiC showed electronic properties comparable to that of isolated graphene sheet [39]. However, the charge transport properties were lower than the pristine graphene by one order. Also, there exists a mixed opinion regarding the bandgap opening in large-area epitaxial graphene [39]. Some reports suggested that a zero band gap in the graphene layer is just above the carbon buffer layer. On the other hand, there are also reports that suggested the zero band gap was found around 0.26 eV [39,63–65]. De Heer et al. [66] developed a method to grow epitaxial graphene on SiC substrate. This study reported that the mobility of graphene grown on a carbon-terminated face was greater than the graphene grown on a silicon-terminated face due to the difference in the structure, and this can be gated [39]. In another work, De Heer et al. [64] evidenced that the energy gap of about 0.26 eV decreased with increasing thickness and approached zero when the number of layers exceeded four. The energy gap was found to be 0.26 eV in single-layer and 0.14 eV in bilayer and triple-layer graphene, respectively, as a result of symmetry breaking caused by the interaction with the substrate.

Emtsev and coworkers [67] achieved high electronic mobilities (2000 cm²/Vs at 27 K and 930 cm²/Vs at 300 K) in a silicon-terminated face in wafer-size graphene layers. Wafer-size graphene layers were prepared in ex situ atmospheric pressure graphitization of SiC in argon atmosphere [39]. However, in another study the mobility variation for multilayer epitaxial graphene films was reported to be 600 to 1200 cm^2/Vs for a silicon face measured in ambient conditions, and it reached 5000 cm^2/Vs in the carbon-terminated face with on/off ratio up to seven [39]. The observed difference was attributed to the different crystallographic domain size of graphene on the two faces [39,61]. The significantly larger single-crystalline domain size was reported for C-face graphene than silicon-face graphene [39,68], which provides structural coherency for a longer mean free path to charges and thus provides high mobility.

For the bulk synthesis of graphene, the CVD technique is often employed for the large-scale synthesis of graphene layers. Monolayer and few-layer growth on various substrates has been demonstrated and discussed. Reina et al. [69], evidenced SLG and few-layer graphene (20-μm lateral size) growth at ambient pressure on nickel substrate by employing the CVD technique. However, the nonhomogeneous thickness of graphene films and grain boundary scattering inside the film caused ineffective modulation; thereby, a huge variation in the field effect mobility (~100–2000 cm^2/Vs) has been witnessed for both electrons and holes [39]. The charge mobility of large-area graphene film grown on nickel by the CVD technique followed by transfer onto the SiO_2 substrate was found to be greater than 3700 cm^2/Vs, which exhibited the half-integer QHE identical to a micromechanically cleaved layer that corresponds to the monolayer characteristic of graphene. Nickel is the widely preferred substrate to grow large-area, high-quality mono- and few-layer graphene [39,70–74]. However, graphene grown on this substrate suffers from small grain size, multilayer deposition at grain boundaries, and carbon solubility in nickel [39].

In another work, Ruoff and coworkers [60] employed a copper foil (as carbon has low solubility in copper) to grow SLG in larger proportions; their result was less than 5% of the area covered with few-layer graphene. The films have achieved mobility as high as 4050 cm²/Vs. Zhou et al. [71] also revealed the potential of the CVD technique to deposit graphene film on a large wafer (up to 3 inches) of copper substrate at ambient pressure. This study showed ambipolar field effect mobilities up to about 3000 cm²/Vs with an on/off ratio of about 4 and half-integer QHE for the films.

Difficulties involved in scaling up single-layer graphene for applications have led researchers to extend the study of graphene to bilayer and few-layer (<10) graphene. Bilayer graphene also exhibits an anomalous QHE, different from that of the single layer. The charge carriers in bilayer graphene have a parabolic energy spectrum near the K point, thereby revealing the gapless nature of bilayer graphene (Figure 1.5) [39,74]. Further, bilayer graphene is metallic at the neutrality points [39,74]. In the case of bilayer graphene, the charge particles are chiral, similar to massless Dirac fermions, but have a finite mass (0.05 m_o).

Although bilayer graphene is gapless, its electronic band gap can be controlled by an electric field perpendicular to the plane [39,66]. The double-gated approach was preferred for demonstrating the controlled induction of an insulating state with large suppression of the conductivity in bilayer graphene [39]. The size of the gap is proportional to the voltage drop between the two graphene planes, and its value can be as high as 0.1–0.3 eV. In contrast, a research group at IBM Thomas J. Watson Research Center [75] showed bilayer graphene FET with a high on/off current ratios of around 100 and 2000 at room temperature and 20 K, respectively, in dual-gated bilayer graphene FETs. The measured band gap was more than 0.13 eV [39].

Various other studies revealed that the electronic band gap of bilayer graphene can be tuned by applying an electric field, which provides an opportunity to use bilayer graphene as a tunable energy band gap semiconductor for many electronic applications, such as photodetectors, terahertz technology items, infrared nanophotonics, pseudospintronics, and laser

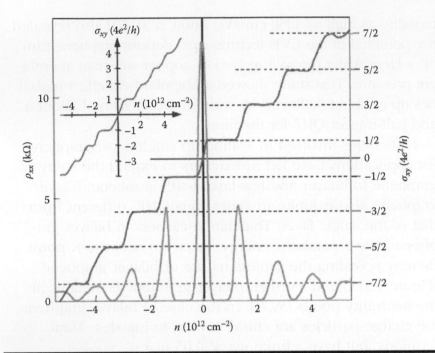

Figure 1.5 (See color insert.)   QHE for massless Dirac Fermions. Hall conductivity σ_{xy} and longitudinal resistivity ρ_{xx} of graphene as a function of their concentration at B = 14T and T = 4K. $\sigma_{xy} \equiv (4e^2/h)\nu$ is calculated from the measured dependencies of $\rho_{xy}(v_g)$ and $\rho_{xx}(v_g)$ as $\rho_{xy} = \rho_{xy}/(\rho^2_{xy} + \rho^2_{xx})$. The behavior of $1/\rho_{xy}$ is similar but exhibits a discontinuity at $v_g \approx 0$, which is avoided by plotting σ_{xy}. Inset: σ_{xy} in two layer graphene where the quantization sequence is normal and occurs at integer v. The latter shows that the half-integer QHE is exclusive to ideal-graphene. (Reproduced with permission from *Nature* 438, 197–200, 2005.)

products [39,76–83]. They have also shown the band gap can be tuned to values larger than 0.2 eV.

Finally, processable graphene sheets in large quantities are also desirable for applications like graphene-reinforced composites, transparent electrical conductive films, energy storage, and so on [39]. A promising approach, including chemical and thermal reduction of graphene oxide (GO), can provide graphene in a scalable amount [26,39,84]. Researchers have been making a continuous effort to improve electrical transport properties of reduced graphene oxide (RGO) for use in electrical applications [39]. Graphene oxide has very high electrical resistance (4 MΩ/square) due to the presence of oxygenated functional groups [39]. Removal of an attached functional group from 2D carbon lattice by chemical and thermal reduction partially restores the electronic conductivity; however, these processes introduce structural imperfections in the carbon lattice and degrade the electrical properties compared to pristine graphene [39]. The conductivity and mobility of RGO were reported to be less by three and two orders of magnitude, respectively, than pure graphene [39]. Lattice vacancies are responsible for reduced conductivity and cannot be healed during the reduction process. The presence of an intact "nanometer"-size domain in RGO creates hopping conduction. Further, field effect mobilities of 2–200 cm^2/Vs and conductivity of 0.05–2 S/cm were reported for RGO [39,78]. Graphene sheets obtained by the hydrazine reduction method function as a P-type semiconductor [39,79].

In contrast, Chhowalla et al. [83] reported that RGO films exhibited ambipolar characters comparable to that of graphene at low temperature. The scattering observed at the junction of the overlapping sheets caused low mobility (1 cm^2/Vs for holes and 0.2 cm^2/Vs for electrons). RGO sheet resistance was found to be as low as 43 KΩ/square prepared by chemical reduction. The lower resistance of 10^2–10^3 Ω/square and electrical conductivity of about 100 S/cm were observed for thermally reduced GO (RGO) [39,81] on comparison with GO (4 MΩ/square) due to high-temperature deoxygenation. Recently, a graphene paper consisting of chemically converted graphene was prepared with electrical conductivity of 72 S/cm at room temperature [39,79]. This graphene paper will have potential applications in membranes, anisotropic conductors, transparent electrodes, and supercapacitors. Refer to the articles by Neto et al. [85] and Nilsson et al. [51] for more information on electronic properties of SLG and bilayer and few-layer graphene.

1.2.2 Quantum Hall Effect

The charge particles behave as massless Dirac fermions in a 2D lattice, which has an interesting effect on the energy spectrum of the Landau levels produced in the presence of a magnetic field perpendicular to the graphene sheets.

The Landau level energies are given by

$$Ej = \frac{\left(j + \frac{1}{2}\hbar eB\right)}{me}$$

where j (= 0, 1, 2, 3, ...) are the indices for the Landau index, and $\hbar = h/2p$.

Because of disorder, the Hall conductivity σ_{xy} of the two-dimensional electron gas (2DEG) shows the plateaus at $jh/2eB$ and is quantized. This is expressed by $\sigma_{xy} = j(2e^2/h)$, and it leads to the IQHE.

In contrast to 2DEG, the energy of Landau levels for graphene is expressed by $Ej = \pm vf\sqrt{2e\hbar B|j|}$. Here $|j| = 0, 1, 2, ...$ is the Landau index, and B is the magnetic field applied perpendicular to the graphene plane. Because the Landau levels are doubly degenerate for the K and K' points at $j = 0$, these levels are equally shared among the electrons and holes. This provides anomalous IQHE, and the Hall conductivity is given by

$$\sigma_{xy} = \pm\frac{4\left(j + \frac{1}{2}\right)e^2}{h}$$

The Hall conductivity for SLG exhibits a plateau when plotted as a function of carrier concentration n at a fixed magnetic field. Novoselov et al. first noticed this anomalous IQHE (shown in Figure 1.5). Further, they noticed a finite value of conductivity ($\sigma_{xy} = 2e^2/h$) at the zero-energy level ($j = 0$). The QHE plateaus occur at the half integer for a high-energy level. This provides a ladder of equidistant steps in the Hall conductivity. In SLG, this QHE is distinctively different for both the holes and electrons compared to conventional QHE as the quantization condition is shifted by a half integer. This observed topologically exceptional electronic structure of graphene was due to abnormal quantization [2,39]. In general, QHE is observed at low temperatures, typically below the boiling point of liquid helium.

Novoselov et al. [3] observed QHE in graphene at room temperature caused by a high concentration of carriers (up to $10^{13}/cm^2$). Moreover, high mobility ($\mu \sim 10,000$ cm^2/Vs) allows the movement of massless Dirac fermions with minimal scattering under ambient conditions. In the case of bilayer graphene, charge carriers have exhibited parabolic energy spectrums that are chiral with finite mass [39,75]. The Landau quantization of these massive Dirac fermions results in plateaus in Hall conductivity that appear at standard integer positions. The Hall conductivity is given by $\sigma_{xy} = j\ 4e^2/h$, where j is an integer other than zero. The first plateau occurs at $j = 1$. However, the plateau at zero rxy is absent (unlike conventional QHE). The Hall conductivity undergoes a double-size step across this region.

1.2.3 Optical Properties

Several reports confirmed that SLG absorbs 2.3% of incident light over a broad wavelength range (Figure 1.6). Graphene transmittance can be well described in terms of fine-structure constants [38,39,86]. The absorption of light was found to increase linearly on the addition of a number of layers (each layer absorption $A = 1 - T = \pi\alpha = 2.3\%$, where $\alpha = 1/37$ is the fine-structure constant). Graphene can be imaged by optical image contrast on Si/SiO$_2$ substrate due to interference, and the contrast increases with the number of layers.

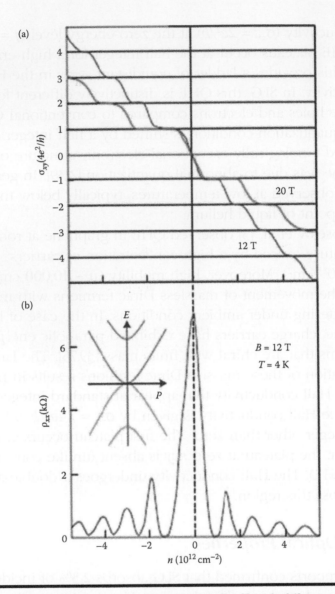

Figure 1.6 (See color insert.) Quantum Hall effect in bilayer graphene. (a) Hall conductivity σ_{xy} and (b) longitudinal ρ_{xx} are plotted
as functions of n at a fixed B and temperature $T = 4$ K. σ_{xy} allows the
underlying sequences of QHE plateaus to be seen more clearly. rxy
crosses zero without any sign of the zero-level plateau as expected in a
conventional 2D system. The inset shows the calculated energy spectrum for bilayer graphene, which is parabolic at low e. (Reproduced
with permission from K.S. Novoselov, E. McCann, S.V. Morozov, V.I.
Fal'ko, M.I. Katsnelson, U. Zeitler, et al., *Nat. Phys.* 2, 177–180, 2006.)

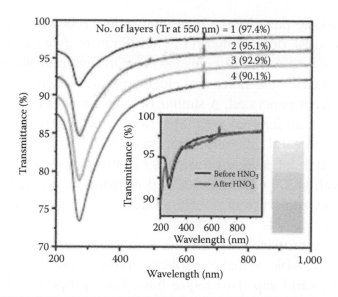

Figure 1.7 (See color insert.)   Representative of transmittance of different graphene layers. UV-vis spectra roll-to-roll, layer-by-layer transferred graphene films on quartz substrates. The inset shows the UV spectra of graphene films with and without HNO$_3$ doping. (Reproduced with permission from S. Bae, H. Kim, Y. Lee, X. Xu, J.-S. Park, Y. Zheng, et al., *Nat. Nanotechnol.* 5, 574–578, 2010.)

The absorption for monolayer graphene is flat from 300 to 2500 nm; the peak at about 250 nm in the ultraviolet (UV) region is attributed to the interband electronic transition from the unoccupied p-states (Figure 1.7) [39,86]. In addition, their optical transition can be altered by varying the Fermi energy effectively through electrical gating [29,39,85]. The tunability, provided by the electrical gating and charge injection in graphene-based optoelectronics devices, has been identified to develop tunable infrared (IR) detectors, modulators, and emitters [29,39]. The exceptional electrical transport properties in conjunction with optical properties have propelled much interest in novel photonic devices. It has also been suggested that the zero band gap, large-area monolayer, and few-layer graphene FETs can be used as ultrafast photodetectors [39,87]. The absorption of light on the surface generates electron–hole pairs in graphene that would recombine quickly (picoseconds), depending on the temperature as well as the density of electrons and holes [39,88]. On the application of an external field, these holes and electrons can be separated and photocurrent generated. A similar behavior was noticed in the presence of an internal field. This field will be generated near the electrode and graphene interface [39,87–90]. Bae et al. [87] showed that the unique properties of graphene provide high-bandwidth (>500 GHz) light detection, wide wavelength detection range, zero current operation, and excellent quantum efficiency.

Another property of graphene is its photoluminescence (PL) [39]. It is possible to make graphene luminescent by inducing a suitable band gap. Two routes have been proposed; the first one involves cutting graphene in nanoribbons and quantum dots [39]. The second one is the physical or chemical treatment of graphene with various gases to reduce the connectivity of the pi-electron network [39,91–93]. For example, it is shown that PL can be induced by oxygen plasma treatment of a graphene single layer on substrate [39,94]. This provides an opportunity to design or create hybrid structures by etching just the top layer, keeping the underlying layer unaffected. Broad PL from solid GO and liquid GO suspension was also observed, and the progressive chemical reduction quenched the PL of GO. Oxidation disrupted the pi-network and opened a direct electronic band gap [39,95].

Fluorescent organic compounds are important for the development of low-cost optoelectronic devices [39,96]. Particularly, blue fluorescence associated with aromatic or olefinic molecules and their derivatives are important in display and lighting applications [39,97]. For example, a thin film of GO deposited from exfoliated suspensions exhibited blue PL [39,98]. PL characteristics and its dependence on the reduction of GO results from the recombination of electron–hole pairs localized within small sp^2 carbon clusters embedded within the GO sp^3 matrix [39,99].

The combined optical and electrical properties of graphene opened new avenues for various photonics and optoelectronics applications. Other exciting and possible applications of graphene include use as photodetectors, touch screens, light-emitting devices, photovoltaics, transparent conductors, terahertz devices, and optical limiters [39].

1.2.4 Mechanical Properties

Unwanted strain may ruin the performance and lifetime of electronic devices. Usually, application of external stress on crystalline material will vary the interatomic distances and cause redistribution and local electronic charge. This results in the generation of a band gap in the electronic structure and significantly affects the electron transport property. Graphene has been reported to have the highest elastic modulus and strength. A single defect-free graphene layer is expected to show the highest intrinsic tensile strength, with stiffness similar to graphite. A method to determine the intrinsic mechanical properties is to investigate the phonon frequency variation on the application of tensile and compressive stress [39,100–104]. Raman spectroscopy is one of the techniques capable of monitoring phonon frequency under uniaxial tensile and hydrostatic stress [39,100–104]. Tensile stress usually induces phonon softening as a consequence of decreased vibrational frequency. On the other hand, compressive stress (hydrostatic) causes phonon hardening due to an increased vibrational frequency mode [39]. As a result, in graphene, investigating the vibration of phonon frequency as a function of strain is expected to provide useful information on stress transfer to individual bonds (for suspended graphene) and atomic-level interaction of graphene to the underlying substrate (for supported graphene) [39].

Compressive and tensile strain in a graphene layer was estimated using Raman spectroscopy by noticing the change in the G and 2D peaks with the applied stress. Increase in the strain causes the "G" peak split and a red shift [39]. For very small strains of about 0.8%, a 2D split was noticed without any shoulder [100–104]. On the other hand, Ni et al. noticed contradicting behavior for epitaxial graphene grown on the SiC substrate [105]. In their work, Ni et al. [105] noticed a blue shift in all the Raman bands for the epitaxial graphene on comparison with the micromechanically cleaved graphene due to the compressive stress associated with the graphene grown thermally. Moreover, the strain on graphene may alter the electronic band structure, indicating that the energy gap can be tuned by introduction of the controlled strain. Band gap tuning was reported under uniaxial strain [39,103,104]. To tune the band gap, a single layer of graphene was deposited on flexible polyethylene terephthalate (PET) so that uniaxial tensile strain (up to ~0.8%) can be applied on SLG/three-layer graphene by stretching the PET in one direction. A band gap of 0.25 eV was detected under the highest strain (0.78%) for the SLG. The uniaxial strain also affected the electronic properties of graphene more significantly as it broke the bonds of the CAC lattice.

1.2.5 Thermal Properties

Thermal properties of graphene play a key role for the application of graphene in electronic devices [39]. Thermal management is also an important factor that contributes to better performance and reliability of electronic components [39]. The amount of heat generated during device operation has to be dissipated. Carbon allotropes such as graphite, diamond, and carbon nanotubes usually exhibit higher thermal conductivity due to strong CAC covalent bonds and phonon scattering [39]. Of all carbon allotropes, carbon nanotubes exhibit the highest thermal conductivity. At room room temperature, multiwall carbon nanotubes (MWCNTs) and single-wall carbon nanotubes (SWCNTs) exhibit a thermal conductivity value of about 3000 W/mK and 3500 W/mK, respectively [39,106–108].

However, a large thermal contact resistance is the major issue with CNT-based semiconductors. In a recent report, defect-free graphene exhibited the highest room temperature thermal conductivity (5000 W/mK) [6,39]. In the case of supported graphene, the conductivity was about 600 W/mK.

Conductivity of graphene on various supports has not been studied in detail. However, its effect was predicted by Klemens [109]. A novel strategy to determine the thermal conductivity of a thin atomic layer of graphene is shown in Figure 1.8 [110]. In this method, suspensions of graphene layer will be heated by laser light (488 nm). As a result, the heat will propagate laterally toward the sinks at the corner of the flakes. The temperature change was determined by measuring the shift in the graphene G peak using confocal micro-Raman spectroscopy, which acts as a thermometer. The thermal conductivity is affected by factors such as defect edge scattering and isotopic doping [39,110–112]. In conclusion, all these factors contribute to the conductivity as a result of phonon scattering at a defect and phonon mode localization due to doping.

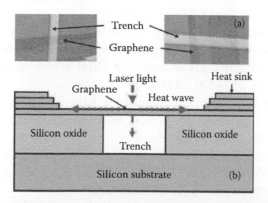

Figure 1.8 (See color insert.)   (a) High-resolution scanning electron microscopic image of the suspended graphene flakes. (b) Schematic of the experimental setup for measuring the thermal conductivity of graphene. (Reproduced with permission from S. Ghosh, I. Calizo, D. Teweldebrhan, E.P. Pokatilov, D.L. Nika, A.A. Balandin, et al., *Appl. Phys. Lett.* 92, 151911, 2008.)

References

1. P.R. Wallace, The band theory of graphite, *Phys. Rev.* 71, 622–634 (1947).
2. J.W. McClure, Diamagnetism of graphite, *Phys. Rev.* 104, 666–671 (1956).
3. J.C. Slonczewski, P.R. Weiss, Band structure of graphite, *Phys. Rev.* 109, 272–279 (1958).
4. G.W. Semenoff, Condensed-matter simulation of a three-dimensional anomaly, *Phys. Rev. Lett* 53, 2449–2452 (1984).
5. E. Fradkin, Critical behavior of disordered degenerate semiconductors, *Phys. Rev. B* 33, 3263–3268 (1986).
6. F.D.M. Haldane, Model for a quantum Hall effect without Landau levels: Condensed-matter realization of the "parity anomaly," *Phys. Rev. Lett.* 61, 2015–2018 (1988).
7. A.K. Geim, K.S. Novoselov, The rise of graphene, *Nat. Mater.* 6, 183–191 (2007).
8. K.S. Novoselov, A.K. Geim, S.V. Morozov, D. Jiang, Y. Zhang, S.V. Dubonos, I.V. Grigorieva, A.A. Firsov, Electric field effect in atomically thin carbon films, *Science* 306, 666–669 (2004).
9. D. Jiang, F. Schedin, T.J. Booth, V.V. Khotkevich, S.V. Morozov, A.K. Geim, Two-dimensional atomic crystals, *Proc. Natl. Acad. Sci. USA* 102, 10451–10453 (2005).
10. K.S. Novoselov, A.K. Geim, S.V. Morozov, D. Jiang, M.I. Katsnelson, I.V. Grigorieva, S.V. Dubonos, A.A. Firsov, Two-dimensional gas of massless Dirac fermions in graphene, *Nature* 438, 197–200 (2005).
11. Y. Zhang, J.W. Tan, H.L. Stormer, P. Kim, Experimental observation of the quantum Hall effect and Berry's phase in graphene, *Nature* 438, 201–204 (2005).
12. G. Brumfiel, Andre Geim: In Praise of Graphene, October 7, doi:10.1038/news.2010.525 (2010). http://www.nature.com/news/2010/101007/full/news.2010.525.html.
13. Associated Press, Graphene Pioneers Earn Nobel Prize in Physics, October 5, 2010. http://www.foxnews.com/scitech/2010/10/05/uk-nobel-prize-physics-graphene/.
14. CNN Wire Staff, Research into Graphene Wins Nobel Prize, October 5, 2010. http://www.cnn.com/2010/LIVING/10/05/sweden.nobel.physics/index.html.

15. B. Partoens, F.M. Peeters, From graphene to graphite: Electronic structure around the K point, *Phys. Rev. B* 74, 075404 (2006).
16. S.V. Morozov, K.S. Novoselov, F. Schedin, D. Jiang, A.A. Firsov, A.K. Geim, Two-dimensional electron and hole gases at the surface of graphite, *Phys. Rev. B* 72, 201401 (2005).
17. M.S. Dresselhaus, G. Dresselhaus, Intercalation compounds of graphite, *Adv. Phys.* 51, 1–186 (2002).
18. H. Shioyama, Cleavage of graphite to graphene, *J. Mater. Sci. Lett.* 20, 499–500 (2001).
19. L.M. Viculis, J.J. Mack, R.B. Kaner, A chemical route to carbon nanoscrolls, *Science* 299, 1361 (2003).
20. S. Mazzocchi, Five Things You Need to Know about the Big Little Substance Graphene, October 8, 2010. http://www.pbs.org/wnet/need-to-know/five-things/the-big-little-substance-graphene/4146/.
21. C. Lee, X.Wei, J.W. Kysar, J. Hone, Measurement of the elastic properties and intrinsic strength of monolayer graphene, *Science* 321, 385–388 (2008).
22. A.A. Balandin, S. Ghosh, W. Bao, I. Calizo, D. Teweldebrhan, F. Miao, C.N. Lau, Superior thermal conductivity of single-layer graphene, *Nano. Lett.* 8, 902–907 (2008).
23. K.I. Bolotin, K.J. Sikes, Z. Jiang, M. Klima, G. Fudenberg, J. Hone, P. Kim, H.L. Stormer, Ultrahigh electron mobility in suspended graphene, *Solid State Commun.* 146, 351–355 (2008).
24. M.D. Stoller, S. Park, Y. Zhu, J. An, R.S. Ruoff, Graphene-based ultracapacitors, *Nano Lett.* 8, 3498–3502 (2008).
25. Y. Zhang, Y.-W. Tan, H.L. Stormer, P. Kim, Experimental observation of the quantum Hall effect and Berry's phase in graphene, *Nature* 438, 201–204 (2005).
26. M.J. Allen, V.C. Tung, R.B. Kaner, Honey comb graphene: A review of graphene, *Chem. Rev.* 110, 132 (2010).
27. S. Alwarappan, A. Erdem, C. Liu, C.-Z. Li, Probing the electrochemical properties of graphene nanosheets for biosensing applications, *J. Phys. Chem. C* 113, 8853–8857 (2009).
28. S. Alwarappan, C. Liu, A. Kumar, C.-Z. Li, Enzyme-doped graphene nanosheets for enhanced glucose biosensing, *J. Phys. Chem. C* 114, 12920–12924 (2010).
29. Y. Liu, D. Yu, C. Zeng, Z.-C. Miao, L. Dai, Biocompatible graphene oxide-based glucose biosensors, *Langmuir* 26, 6158–6160 (2010).

30. M. Zhou, Y.M. Zhai, S.J. Dong, Electrochemical biosensing based on reduced graphene oxide, *Anal. Chem.* 81, 5603–5613 (2009).

31. D.A. Dikin, Preparation and characterization of graphene oxide paper, *Nature* 448, 457–460 (2007).

32. S. Park, K.S. Lee, G. Bozoklu, W. Cai, S.T. Nguyen, R.S. Ruoff, Graphene oxide papers modified by divalent ions-enhancing mechanical properties *via* chemical cross-linking, *ACS Nano* 2, 572–578 (2008).

33. S. Stankovich, D.A. Dikin, G.H.B. Dommet, K.M. Kohlhaas, E.J. Zimney, E.A. Stach, R.D. Piner, S.T. Nguyen, R. Ruoff, Graphene-based composite materials, *Nature* 442, 282–286 (2006).

34. T. Ramanathan, A.A. Abdala, S. Stankovich, D.A. Dikin, M. Herrera-Alonso, R.D. Piner, D.H. Adamson, H.C. Schniepp, X. Chen, R.S. Ruoff, S.T. Nguyen, I.A. Aksay, R.K. Prud'Homme, L.C. Brinson, Functionalized graphene sheets for polymer nanocomposites, *Nat. Nanotechnol.* 3, 327–331 (2008).

35. P. Blake, P.D. Brimicombe, R.R. Nair, T.J. Booth, D. Jiang, F. Schedin, L.A. Ponomarenko, S.V. Morozov, H.F. Gleeson, E.W. Hill, A.K. Geim, K.S. Novoselov, Graphene-based liquid crystal device, *Nano Lett.* 8, 1704–1708 (2008).

36. J.S. Bunch, A.M. Van der Zande, S.S. Verbridge, I.W. Frank, D.M. Tanenbaum, J.M. Parpia, G.H. Craighead, P.L. McEuen, Electromechanical resonators from graphene sheets, *Science* 315, 490–493 (2007).

37. K.I. Bolotin, K.J. Sikes, Z. Jiang, M. Klima, G. Fudenberg, J. Hone, et al., Ultrahigh electron mobility in suspended graphene, *Solid State Commun.* 146, 351–355 (2008).

38. R.R. Nair, P. Blake, A.N. Grigorenko, K.S. Novoselov, T.J. Booth, T. Stauber, et al., Fine structure constant defines visual transparency of graphene, *Science* 320, 1308 (2008).

39. V. Singh, D. Joung, L. Zhai, S. Das, S.I. Khondaker, S. Seal, Graphene based materials: Past, present and future, *Prog. Mater. Sci.* 56, 1178–1271 (2011).

40. T.B. Zhang, Y.W. Tan, H.L. Stormer, P. Kim, Experimental observation of the quantum Hall effect and Berry's phase in graphene, *Nature* 438, 201–204 (2005).

41. K.S. Novoselov, D. Jiang, F. Schedin, T.J. Booth, V.V. Khotkevich, S.V. Morozov, et al., Two-dimensional atomic crystals, *Proc. Natl. Acad. Sci. USA* 102, 10451–10453 (2005).

42. K.S. Novoselov, Z. Jiang, Y. Zhang, S.V. Morozov, H.L. Stormer, U. Zeitler, et al., Room-temperature quantum Hall effect in graphene, *Science* 315, 1379 (2007).
43. K. Nomura, A.H. MacDonald, Quantum Hall ferromagnetism in graphene, *Phys. Rev. Lett.* 96, 256602 (2006).
44. E.H. Hwang, S. Adam, S. Das Sarma, Carrier transport in two-dimensional graphene layers, *Phys. Rev. Lett.* 98, 186806 (2007).
45. J.C. Meyer, A.K. Geim, M.I. Katsnelson, K.S. Novoselov, T.J. Booth, S. Roth, The structure of suspended graphene sheets, *Nature* 446, 60–63 (2007).
46. Y.-W. Son, M.L. Cohen, S.G. Louie, Energy gaps in graphene nanoribbons, *Phys. Rev. Lett.* 97, 216803 (2006).
47. M.Y. Han, B. Ozyilmaz, Y. Zhang, P. Kim, Energy band-gap engineering of graphene nanoribbons, *Phys. Rev. Lett.* 98, 206805 (2007).
48. Z. Chen, Y.-M. Lin, M.J. Rooks, P. Avouris, Graphene nano-ribbon electronics, *Phys. E. Low-Dimen. Syst. Nanostruct.* 40, 228–232 (2007).
49. B. Trauzettel, D.V. Bulaev, D. Loss, G. Burkard, Spin qubits in graphene quantum dots, *Nat. Phys.* 3, 192–196 (2007).
50. T. Ohta, A. Bostwick, T. Seyller, K. Horn, E. Rotenberg, Controlling the electronic structure of bilayer graphene, *Science* 313, 951–954 (2006).
51. J. Nilsson, A.H. Castro Neto, F. Guinea, N.M.R. Peres, Electronic properties of bilayer and multilayer graphene, *Phys. Rev. B* 78, 045405 (2008).
52. Y. Zhang, T.-T. Tang, C. Girit, Z. Hao, M.C. Martin, A. Zettl, et al., Direct observation of a widely tunable bandgap in bilayer graphene, *Nature* 459, 820–823 (2009).
53. M. Evaldsson, I.V. Zozoulenko, H. Xu, T. Heinzel, Edge-disorder-induced Anderson localization and conduction gap in graphene nanoribbons, *Phys. Rev. B* 78, 161407 (2008).
54. F. Cervantes-Sodi, G. Csanyi, S. Piscanec, A.C. Ferrari, Edge-functionalized and substitutionally doped graphene nanoribbons: Electronic and spin properties, *Phys. Rev. B* 77, 165427 (2008).
55. M.C. Lemme, T.J. Echtermeyer, M. Baus, H. Kurz, A graphene field-effect device, *IEEE Electron. Dev. Lett.* 28, 282–284 (2007).
56. L. Liao, J. Bai, Y. Qu, Y.-C. Lin, Y. Li, Y. Huang, et al., High j oxide nanoribbons as gate dielectrics for high mobility top-gated graphene transistors, *Proc. Natl. Acad. Sci. USA* 107, 6711–6715 (2010).

57. Y.-M. Lin, K.A. Jenkins, A. Valdes-Garcia, J.P. Small, D.B. Farmer, P. Avouris, Operation of graphene transistors at gigahertz frequencies, *Nano Lett.* 9, 422–426 (2008).

58. L. Liao, J. Bai, R. Cheng, Y.-C. Lin, S. Jiang, Y. Huang, et al., Top-gated graphene nanoribbon transistors with ultrathin high-k dielectrics, *Nano Lett.* 10, 1917–1921 (2010).

59. J. Kedzierski, P.-L. Hsu, P. Healey, P.W. Wyatt, C.L. Keast, M. Sprinkle, et al., Epitaxial graphene transistors on SiC substrates, *IEEE Trans. Electron. Dev.* 55, 2078–2085 (2008).

60. X. Li, W. Cai, L. Colombo, R.S. Ruoff, Evolution of graphene growth on Ni and Cu by carbon isotope labeling, *Nano Lett.* 9, 4268–4272 (2009).

61. X. Peng, R. Ahuja, Symmetry breaking induced bandgap in epitaxial graphene layers on SiC, *Nano Lett.* 8, 4464–4468 (2008).

62. S.Y. Zhou, G.H. Gweon, A.V. Fedorov, P.N. First, W.A. De Heer, D.H. Lee, et al., Substrate-induced bandgap opening in epitaxial graphene, *Nat. Mater.* 6, 770–775 (2007).

63. S. Kim, J. Ihm, H.J. Choi, Y.-W. Son, Origin of anomalous electronic structures of epitaxial graphene on silicon carbide, *Phys. Rev. Lett.* 100, 176802 (2008).

64. W.A. De Heer, C. Berger, X. Wu, P.N. First, E.H. Conrad, X. Li, et al., Epitaxial graphene, *Solid State Commun.* 143, 92–100 (2007).

65. J.B. Oostinga, H.B. Heersche, X. Liu, A.F. Morpurgo, L.M.K. Vandersypen, Gate-induced insulating state in bilayer graphene devices, *Nat. Mater.* 7, 151–157 (2008).

66. J. Hass, R. Feng, T. Li, X. Li, Z. Zong, W.A. De Heer, et al., Highly ordered graphene for two dimensional electronics, *Appl. Phys. Lett.* 89, 143106 (2006).

67. K.V. Emtsev, A. Bostwick, K. Horn, J. Jobst, G.L. Kellogg, L. Ley, et al., Towards wafer-size graphene layers by atmospheric pressure graphitization of silicon carbide, *Nat. Mater.* 8, 203–207 (2009).

68. K.S. Kim, Y. Zhao, H. Jang, S.Y. Lee, J.M. Kim, K.S. Kim, et al., Large-scale pattern growth of graphene films for stretchable transparent electrodes, *Nature* 457, 706–710 (2009).

69. A. Reina, X. Jia, J. Ho, D. Nezich, H. Son, V. Bulovic, et al., Large area, few-layer graphene films on arbitrary substrates by chemical vapor deposition, *Nano Lett.* 9, 30–35 (2008).

70. H. Cao, Q. Yu, R. Colby, D. Pandey, C.S. Park, J. Lian, et al., Large-scale graphitic thin films synthesized on Ni and transferred to insulators: Structural and electronic properties, *J. Appl. Phys.* 107, 044310 (2010).
71. L.G. Arco, Y. Zhang, A. Kumar, C. Zhou, Synthesis, transfer, and devices of single- and few-layer graphene by chemical vapor deposition, *IEEE Trans. Nanotechnol.* 8, 135–138 (2009).
72. Q. Yu, J. Lian, S. Siriponglert, H. Li, Y.P. Chen, S.-S. Pei, Graphene segregated on Ni surfaces and transferred to insulators, *Appl. Phys. Lett.* 93, 113103 (2008).
73. X. Li, W. Cai, J. An, S. Kim, J. Nah, D. Yang, et al., Large-area synthesis of high-quality and uniform graphene films on copper foils, *Science* 324, 1312–1314 (2009).
74. K.S. Novoselov, E. McCann, S.V. Morozov, V.I. Fal'ko, M.I. Katsnelson, U. Zeitler, et al., Unconventional quantum Hall effect and Berry's phase of 2 pi in bilayer graphene, *Nat. Phys.* 2, 177–180 (2006).
75. F. Xia, D.B. Farmer, Y.-M. Lin, P. Avouris, Graphene field-effect transistors with high on/off current ratio and large transport band gap at room temperature, *Nano Lett.* 10, 715–718 (2010).
76. M. Tonouchi, Cutting-edge terahertz technology, *Nat. Photon.* 1, 97 (2007).
77. F. Wang, Y. Zhang, C. Tian, C. Girit, A. Zettl, M. Crommie, et al., Gate-variable optical transitions in graphene, *Science* 320, 206–209 (2008).
78. P. San-Jose, E. Prada, E. McCann, H. Schomerus, Pseudospin valve in bilayer graphene: Towards graphene-based pseudo-spintronics, *Phys. Rev. Lett.* 102, 247204 (2009).
79. D. Li, M.B. Muller, S. Gilje, R.B. Kaner, G.G. Wallace, Processable aqueous dispersions of graphene nanosheets, *Nat. Nanotechnol.* 3, 101–105 (2008).
80. S. Stankovich, D.A. Dikin, R.D. Piner, K.A. Kohlhaas, A. Kleinhammes, Y. Jia, et al., Synthesis of graphene-based nanosheets *via* chemical reduction of exfoliated graphite oxide, *Carbon* 45, 1558–1565 (2007).
81. C. Gomez-Navarro, R.T. Weitz, A.M. Bittner, M. Scolari, A. Mews, M. Burghard, et al., Electronic transport properties of individual chemically reduced graphene oxide sheets, *Nano Lett.* 7, 3499–3503 (2007).

82. S. Gilje, S. Han, M. Wang, K.L. Wang, R.B. Kaner, A chemical route to graphene for device applications, *Nano Lett.* 7, 3394–3398 (2007).
83. G. Eda, G. Fanchini, M. Chhowalla, Large-area ultrathin films of reduced graphene oxide as a transparent and flexible electronic material, *Nat. Nanotechnol.* 3, 270–274 (2008).
84. E.V. Castro, K.S. Novoselov, S.V. Morozov, N.M.R. Peres, J.M.B.L. Dos Santos, J. Nilsson, et al., Biased bilayer graphene: semiconductor with a gap tunable by the electric field effect, *Phys. Rev. Lett.* 99, 216802 (2007).
85. A.H. Castro Neto, F. Guinea, N.M.R. Peres, K.S. Novoselov, A.K. Geim, The electronic properties of graphene, *Rev. Mod. Phys.* 81, 109–162 (2009).
86. H.A. Becerril, J. Man, Z. Liu, R.M. Stoltenberg, Z. Bao, Y. Chen, Evaluation of solution-processed reduced graphene oxide films as transparent conductors, *ACS Nano* 2, 463–470 (2008).
87. S. Bae, H. Kim, Y. Lee, X. Xu, J.-S. Park, Y. Zheng, et al., Roll-to-roll production of 30-inch graphene films for transparent electrodes, *Nat. Nanotechnol.* 5, 574–578 (2010).
88. V.G. Kravets, A.N. Grigorenko, R.R. Nair, P. Blake, S. Anissimova, K.S. Novoselov, et al., Spectroscopic ellipsometry of graphene and an exciton-shifted van Hove peak in absorption, *Phys. Rev. B* 81, 155413 (2010).
89. Z.Q. Li, E.A. Henriksen, Z. Jiang, Z. Hao, M.C. Martin, P. Kim, et al., Dirac charge dynamics in graphene by infrared spectroscopy, *Nat. Phys.* 4, 532–535 (2008).
90. F. Xia, T. Mueller, R. Golizadeh-Mojarad, M. Freitag, Y.M. Lin, J. Tsang, et al., Photocurrent imaging and efficient photon detection in a graphene transistor, *Nano Lett.* 9, 1039–1044 (2009).
91. F. Rana, P.A. George, J.H. Strait, J. Dawlaty, S. Shivaraman, M. Chandrashekhar, et al. Carrier recombination and generation rates for intravalley and intervalley phonon scattering in graphene, *Phys. Rev. B* 79, 115447 (2009).
92. T. Mueller, F. Xia, M. Freitag, J. Tsang, P. Avouris, Role of contacts in graphene transistors: A scanning photocurrent study, *Phys. Rev. B* 79, 245430 (2009).
93. J.H. Lee Eduardo, K. Balasubramanian, R.T. Weitz, M. Burghard, K. Kern, Contact and edge effects in graphene devices, *Nat. Nanotechnol.* 3, 486–490 (2008).
94. S. Park, R.S. Ruoff, Chemical methods for the production of graphenes, *Nat. Nanotechnol.* 4, 217–224 (2009).

95. D.C. Elias, R.R. Nair, T.M.G. Mohiuddin, S.V. Morozov, P. Blake, M.P. Halsall, et al., Control of graphene's properties by reversible hydrogenation: evidence for graphane, *Science* 323, 610–613 (2009).

96. F. Bonaccorso, Z. Sun, T. Hasan, A.C. Ferrari, Graphene photonics and optoelectronics, *Nat. Photon.* 4, 611–622 (2010).

97. T. Gokus, R.R. Nair, A. Bonetti, M. Bohmler, A. Lombardo, K.S. Novoselov, et al., Making graphene luminescent by oxygen plasma treatment, *ACS Nano* 3, 3963–3968 (2009).

98. Z. Luo, P.M. Vora, E.J. Mele, A.T.C. Johnson, J.M. Kikkawa, Photoluminescence and band gap modulation in graphene oxide, *Appl. Phys. Lett.* 94, 111909 (2009).

99. J.R. Sheats, H. Antoniadis, M. Hueschen, W. Leonard, J. Miller, R. Moon, et al., Organic electroluminescent devices, *Science* 273, 884–888 (1996).

100. L.J. Rothberg, A.J. Lovinger, Status of and prospects for organic electroluminescence, *J. Mater. Res.* 11, 3174–3187 (1996).

101. G. Eda, Y.-Y. Lin, C. Mattevi, H. Yamaguchi, H.A. Chen, I.S. Chen, et al., Blue photoluminescence from chemically derived graphene oxide, *Adv. Mater.* 22, 505–509 (2010).

102. T. Yu, Z. Ni, C. Du, Y. You, Y. Wang, Z. Shen, Raman mapping investigation of graphene on transparent flexible substrate: The strain effect, *J. Phys. Chem. C* 112, 12602–12605 (2008).

103. Z.H. Ni, W. Chen, X.F. Fan, J.L. Kuo, T. Yu, A.T.S. Wee, et al., Raman spectroscopy of epitaxial graphene on a SiC substrate, *Phys. Rev. B* 77, 6 (2008).

104. Z.H. Ni, H.M. Wang, Y. Ma, J. Kasim, Y.H. Wu, Z.X. Shen, Tunable stress and controlled thickness modification in graphene by annealing, *ACS Nano* 2, 1033–1039 (2008).

105. Z.H. Ni, T. Yu, Y.H. Lu, Y.Y. Wang, Y.P. Feng, Z.X. Shen, Uniaxial strain on graphene: Raman spectroscopy study and band-gap opening, *ACS Nano* 2, 2301–2305 (2008).

106. T.M.G. Mohiuddin, A. Lombardo, R.R. Nair, A. Bonetti, G. Savini, R. Jalil, et al., Uniaxial strain in graphene by Raman spectroscopy: G peak splitting, Gruneisen parameters, and sample orientation, *Phys. Rev. B* 79, 205433 (2009).

107. P. Kim, L. Shi, A. Majumdar, P.L. McEuen, Thermal transport measurements of individual multiwalled nanotubes, *Phys. Rev. Lett.* 87, 215502 (2001).

108. E. Pop, D. Mann, Q. Wang, K. Goodson, H. Dai, Thermal conductance of an individual single-wall carbon nanotube above room temperature, *Nano Lett.* 6, 96–100 (2005).

109. P.G. Klemens, Theory of thermal conduction in thin ceramic films, *Int. J. Thermophys.* 22, 265–275 (2001).

110. S. Ghosh, I. Calizo, D. Teweldebrhan, E.P. Pokatilov, D.L. Nika, A.A. Balandin, et al., Extremely high thermal conductivity of graphene: prospects for thermal management applications in nanoelectronic circuits, *Appl. Phys. Lett.* 92, 151911 (2008).

111. D.L. Nika, E.P. Pokatilov, A.S. Askerov, A.A. Balandin, Phonon thermal conduction in graphene: Role of Umklapp and edge roughness scattering, *Phys. Rev. B* 79, 155413 (2009).

112. J.-W. Jiang, J. Lan, J.-S. Wang, B. Li, Isotopic effects on the thermal conductivity of graphene nanoribbons: Localization mechanism, *J. Appl. Phys.* 107, 054314 (2010).

Chapter 2

Graphene Synthesis

2.1 Introduction

To date, great effort has been made to form uniform graphene films. Graphene films have been prepared by a variety of techniques, including mechanical exfoliation [1–3], graphene in solution [2–5], and epitaxial growth [2,6,7] methods. Mechanical exfoliation produces the highest-quality graphene, which is suitable for fundamental studies; epitaxial growth provides the shortest path to graphene-based electronic circuits, and graphene in solution can offer lower cost and higher throughput for the production of graphene-based nanocomposites as well as larger-size films than other existing methods [2,8]. This chapter describes these methods in detail.

2.2 Mechanical Exfoliation

For the first time, during 1999, Ruoff's group presented a mechanical exfoliation route for peeling out the graphene planes from graphite using an atomic force microscope (AFM) tip to manipulate small pillars patterned into highly oriented pyrolytic graphite (HOPG) by plasma etching as shown in Figure 2.1 [9].

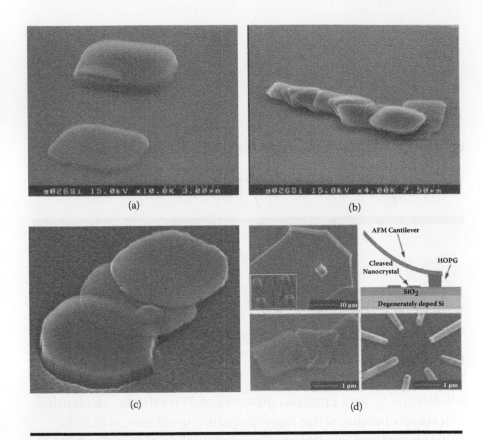

Figure 2.1 Scanning electron microscopic images of early attempts at mechanical exfoliation using graphite pillars. (a) and (b) Ruoff's group peeled away layers with an AFM tip. (Reprinted with permission from X.K. Lu, M.F. Yu, H. Huang, R.S. Ruoff, Tailoring graphite with the goal of achieving single sheets, *Nanotechnology* 10, 269, 1999. Copyright 1999 Institute of Physics.) (c) and (d) Kim's group transferred the pillars to a tipless cantilever and deposited thin slabs onto other substrates in tapping mode. A series of scanning electron microscopic images show thin samples cleaved onto the Si/SiO₂ substrate and a typical mesoscopic device. (Reprinted with permission from Y.B. Zhang, J.P. Small, W.V. Pontius, P. Kim, *Appl. Phys. Lett.* 86, 073104, 2005. Copyright 2005 American Institute of Physics.)

The thinnest slabs observed during that time were more than 200 nm thick or the equivalent of 600 layers.

Later, Kim's group improved the method by transferring the pillars to a tipless cantilever, which successively stamped slabs to about 10 nm (or 30 layers) on SiO_2 [10]. Other early groups working toward peeling graphene from graphite included Enoki et al. [11], who used temperatures around 1600°C to convert nanodiamonds into nanometer regions of graphene. Although all these methods are capable of producing few-layer graphene, it was Geim and Novoselov [8] who introduced a much simpler approach that led to the first isolation of single-layer graphene in 2004 (see Figure 2.2). In its most basic form, the "peeling" method utilizes common cellophane tape to successively remove layers from a graphite flake. The tape is ultimately pressed against a substrate to deposit sample (see Figure 2.3). Although the flakes present on the tape are thicker than a single layer, van der Waals attraction to the substrate can delaminate a single sheet when the tape is then lifted. The method requires a great deal of patience as depositions put down by inexperienced individuals will often result in thicker slabs in which locating a single layer can be extremely difficult. However, after several times (with practice), the technique results in high-quality crystallites, which can be more than 100 μm^2 in size. Perhaps the most important part of isolating single-layer graphene for the first time was the ability to spot an atomically thin specimen in some readily identifiable fashion [12].

Optical absorbance of graphene has since been measured at just 2.3%, ruling out direct visual observation, and is shown in Figure 2.4 [13,14]. To visualize single flakes, Geim and coworkers took advantage of an interference effect at a specially chosen thickness (300 nm) of SiO_2 on silicon to enhance the optical contrast under white light illumination [15]. Although it looks like a simple idea, this was a major step forward and has contributed a great deal toward progress in this field.

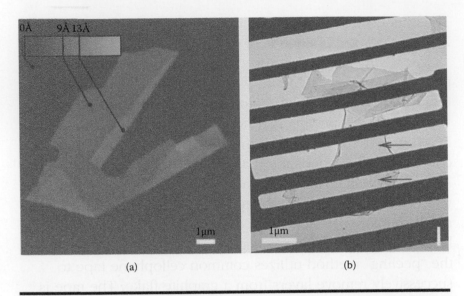

(a)   (b)

Figure 2.2 **Mechanical exfoliation resulted in the first single-layer graphene flakes. (a) An atomic force microscopic image showing the substrate-graphene step height of less than 1 nm and a folded step height of 0.4 nm. (Reproduced with permission from K.S. Novoselov, D. Jiang, F. Schedin, T.J. Booth, V.V. Khotkevich, S.V. Morozov, A.K. Geim,** *Proc. Natl. Acad. Sci. USA* **102, 10451–10453, 2005. Copyright 2005 PNAS.) (b) Transmission electron microscopic (TEM) image of a freestanding graphene film after etching of the underlying substrate. (Reproduced with permission from J.C. Meyer, A.K. Geim, M.I. Katsnelson, K.S. Novoselov, T.J. Booth, S. Roth,** *Nature* **446, 60–63, 2007. Copyright 2007 Nature Publishing Group.)**

2.3 Alternatives to Mechanical Exfoliation

The limiting step for most of the experiments was simply obtaining good single layers of graphene by mechanical exfoliation. This would have greater implications for real-world devices because the process is low throughput and unlikely to be industrially scalable. As a result, the search for an alternate route to synthesize single-layer graphene became the focus of a great deal of research. The other important factors beyond scalability when considering the proficiency of any synthetic route to graphene are the following: (1) A process

Figure 2.3 The single-layer graphene as first observed by Geim et al. at the University of Manchester. Here, a few-layer flake is shown, with optical contrast enhanced by an interference effect at a carefully chosen thickness of oxide. (Reproduced with permission from G. Eda, G. Fanchini, M. Chhowalla, *Nat. Nanotechol.* 3, 270–274, 2008. Copyright 2006 American Association for the Advancement of Science.)

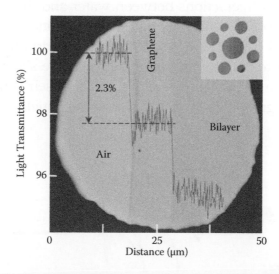

Figure 2.4 A single and bilayer sample suspended on a porous membrane. Optical absorbance was measured at 2.3% per layer. The inset shows the sample design with several apertures. (Reproduced with permission from R.R. Nair, P. Blake, A.N. Grigorenko, K.S. Novoselov, T.J. Booth, T. Stauber, N.M.R. Peres, A.K. Geim, *Science* 320, 1308, 2008. Copyright 2008 American Association for the Advancement of Science.)

must produce high quality in the two-dimensional (2D) crystal lattice to ensure high mobility. (2) The method must provide fine control over crystallite thickness to deliver uniform device performance. (3) For ease of integration, any process should be compatible with current CMOS (complementary metal-oxide semiconductor) processing.

2.3.1 Chemical Method

Ruoff and coworkers demonstrated a solution-assisted process for producing a single layer of graphene as shown in Figure 2.5 [16–18]. In this method, initially graphite will be chemically modified to yield a water-dispersible graphitic oxide (GO) intermediate. The GO thus obtained consists of a layered stack of puckered sheets that completely exfoliates on the addition of mechanical energy [19,20]. This is due to the strength of interactions between water and the oxygen-containing (epoxide and hydroxyl) functionalities introduced into the basal plane during oxidation. The hydrophilicity allows water to readily intercalate between the sheets and disperse them as individuals. Although GO itself is nonconducting, the graphitic network can be substantially restored

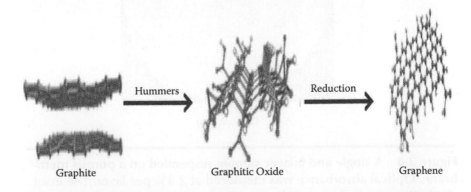

Graphite　　　　　Hummers　→　　　Graphitic Oxide　　　Reduction　→　　Graphene

Figure 2.5 **Molecular models show the conversion process from graphite to chemically derived graphene. (Reproduced with permission from V.C. Tung, M.J. Allen, Y. Yang, R.B. Kaner, *Nat. Nanotechnol.* 4, 25–29, 2009. Copyright 2009 Nature Publishing Group.)**

by thermal annealing or through treatment with chemical reducing agents, a number of which have been explored.

According to Ruoff et al., hydrazine hydrate is the best reagent to eliminate oxidation through the formation and removal of epoxide complexes [16], and this involves the simple addition of hydrazine directly to aqueous dispersions of GO. However, the original aqueous reduction of GO resulted in the removal of oxygen groups; as a result of this, the reduced sheets become less hydrophilic and quickly aggregated in solution. Increasing the pH during the reduction will result in charge stabilized colloidal dispersions, even in the deoxygenated sheets. Tuang et al. improved the reduction step by making dispersions directly in anhydrous hydrazine [21–23]. (Note: The use of hydrazine requires great care because it is both highly toxic and potentially explosive [24].)

The most exciting advantages of the GO method are its low cost and massive scalability. The starting material is simple graphite, and the technique can be easily scaled up to produce larger amounts of larger chemically derived graphene dispersed in a liquid [25,26]. Further, it is important to note that GO is also an interesting material for composite applications, and the tensile strength of freestanding films can be as high as 42 GPa (see Figure 2.6) [27].

2.3.2 *Total Organic Synthesis*

Although chemically derived micrometer-scale graphene was obtained from GO, synthetic techniques for smaller planar, benzene-like macromolecules have been known for a long time [28–33]. These graphene-like polycyclic aromatic hydrocarbons (PAHs) finds an interesting place between "molecular" and "macromolecular" structures and attracts new interest as a possible alternative route to graphene. PAHs are attractive because they are highly versatile and can be substituted with a variety of aliphatic chains to alter solubility [34]. However,

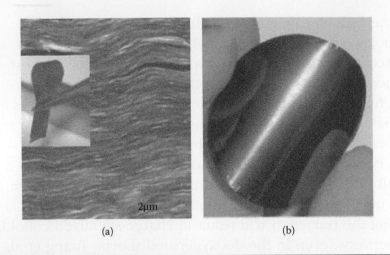

(a)   (b)

Figure 2.6 Freestanding graphene films show extremely high tensile strength. (a) Cross-sectional scanning electron microscopic (SEM) image of graphite oxide stacking in a film produced by filtration. (Reproduced with permission from D.A. Dikin, S. Stankovich, E.J. Zimney, R.D. Piner, G.H.B. Dommett, G. Evmenenko, S.T. Nguyen, R.S. Ruoff, *Nature* 448, 457–460, 2007. Copyright 2007 Nature Publishing Group.) (b) Chemical reduction produces a film with shiny luster. (Reproduced with permission from D. Li, R.B. Kaner, *Science* 320, 1170–1171, 2008. Copyright 2008 American Association for the Advancement of Science.)

the limited size range of PAHs was their major drawback. The limited size of the PAHs can be attributed to the fact that increasing molecular weight generally decreases solubility and increases the occurrence of side reactions. Under these conditions, preservation of dispersibility and a planar morphology for large PAHs will be critical. However, there was a major breakthrough when Mullen and coworkers reported the synthesis of nanoribbon-like PAHs up to 12 nm in length (see Figure 2.7) [29]. Although the electronic properties of these nanoribbons have yet to be characterized, they may indeed exhibit graphene-like behavior. In the future, if researchers in this area can extend the size range of PAHs, this could provide a clean synthetic route to graphene for some applications. In any event, the organic techniques developed will

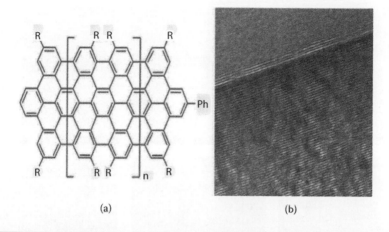

(a)   (b)

Figure 2.7 **Polycyclic aromatic hydrocarbons (PAHs) may offer a ground-up synthesis of graphene. (a) Chemical structure of PAHs and (b) TEM of a nanoribbon synthesized by Mullen. (Reproduced with permission from X.Y. Yang, X. Dou, A. Rouhanipour, L.J. Zhi, H.J. Rader, K.J. Mullen, *J. Am. Chem. Soc.* 130, 4216, 2008. Copyright 2008 American Chemical Society.)**

have important implications for modification of or addition to conjugated carbon macromolecules.

2.3.3 Depositions

2.3.3.1 Overview

Uniform and reproducible depositions are the most important requirements for incorporating a solution-based technique into device fabrication. Furthermore, the required type of deposition can vary widely depending on the design specifics of a given device. Chemically converted graphene suspensions are the best fit for this discussion; they are versatile and permit a variety of deposition techniques (see Figure 2.8) [21,35–37]. These have been used to produce films with coverage ranging anywhere from evenly spaced single sheets to densely packed overlapping films.

The original technique used for depositing films was spray coating from water onto a heated substrate. Although it is

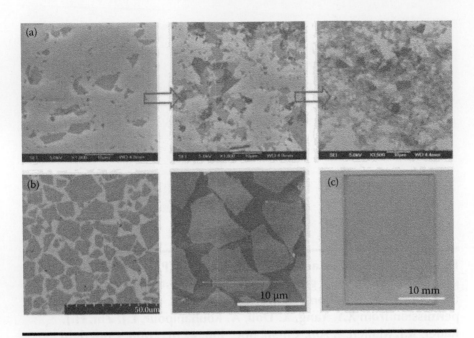

Figure 2.8 Solution processing allows deposition of synthesized/ modified graphene in a variety of densities. (a) SEM images of different films spin coated from hydrazine. (b) SEM and atomic force microscopic images of a graphite oxide film deposited by Langmuir-Blodgett assembly. (Reproduced with permission from L.J. Cote, F. Kim, J. Huang, *J. Am. Chem. Soc.* 131, 1043–1049, 2009. Copyright 2009 American Chemical Society.) (c) Multilayer coatings are still quite transparent. (Reproduced with permission from X. Li, G. Zhang, X. Bai, X. Sun, X. Wang, E. Wang, H. Dai, *Nat. Nanotechnol.* 3, 538–542, 2008. Copyright 2008 Nature Publishing Group.)

possible to isolate and characterize some single sheets, high surface tension caused significant aggregation even if the substrate was heated to flash dry the suspension on contact. Huang and coworkers demonstrated wonderful control over depositions using Langmuir-Blodgett assembly of GO [35]. They showed that electrostatic repulsion prevented the single layers from overlapping when compressed at an air–surface interface. This led to depositions on SiO_2 that included dilute, close-packed, and overpacked films. Dai et al. performed similar work with Langmuir-Blodgett techniques and layer-by-layer assembly using electrostatic attraction to biased substrates [37].

2.3.3.2 Chemical Vapor Deposition

Often, the solution-based synthetic routes were preferred to circumvent the need for support substrates. Of these, two techniques take advantage of specially chosen platforms to encourage the growth of high-quality graphene. Several groups reported an epitaxial method in which graphene was obtained from the high-temperature reduction of silicon carbide (see Figure 2.9) [6,38–42]. Moreover, the process is relatively straightforward as silicon desorbs around 1000°C in an ultrahigh vacuum. This leaves behind small traces of islands of graphitized carbon, which were first identified by scanning tunneling microscopy (STM) and electron diffraction

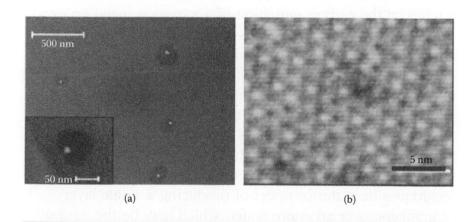

(a)   (b)

Figure 2.9 Silicon carbide is reduced to graphene as silicon sublimes at high temperature. (a) SEM image shows small hexagonal crystallites. (Reproduced with permission from M.L.E.A. Sadowski, *J. Phys. Chem. Solids* 67, 2172–2177, 2006. (b) STM image shows long-range order and a low density of defects. (Reprinted with permission from C. Berger, Z.M. Song, X.B. Li, X.S. Wu, N. Brown, C. Naud, D. Mayou, T.B. Li, J. Hass, A.N. Marchenkov, E.H. Conrad, P.N. First, W.A. De Heer, *Science* 312, 1191, 2006. Copyright 2006 American Association for the Advancement of Science.)

experiments. More recently, there are reports that researchers employed photolithography to pattern epitaxial growth in predetermined locations to make devices [41].

Nevertheless, there are several physical properties that may vary between epitaxially grown and mechanically exfoliated graphene [38,42]. These discrepancies can be attributed to the influence of interfacial effects in epitaxial graphene, which are heavily dependent on both the silicon carbide substrate and several growth parameters. For epitaxial graphene, differences in the periodicity observed by STM and LEED (low-energy electron diffraction) are not exactly defined or understood [43]. The same is true for the energy gap observed by angle-resolved photoemission spectroscopy (ARPES) [44]. On the other hand, the second substrate-based method is chemical vapor deposition (CVD) of graphene on transition metal films (see Figure 2.10). This process was pioneered by researchers at the Massachusetts Institute of Technology (MIT) and in Korea; it relies on the carbon saturation of a transition metal on exposure to a hydrocarbon gas at a very high temperature [45–47]. Most often, nickel films are used along with methane gas. On cooling the substrate, the solubility of carbon in the transition metal decreases, and a thin film of carbon is expected to precipitate from the surface.

One of the major advantages of substrate-based methods for graphene synthesis is their high compatibility with the latest CMOS technology. In theory, both epitaxial and CVD techniques have the prospect of producing a single layer of graphene over an entire wafer, which may be the easiest route to integrate the novel material into the current semiconductor processes and devices. The other challenge that exists in the epitaxial and CVD methods is in obtaining fine control over film thickness and preventing secondary crystal formation. In an ideal case, both the methods rely on the nucleation and growth of a single crystal without the formation of a boundary or seeding of a second layer. Until now, the best specimens have had a variation in thickness

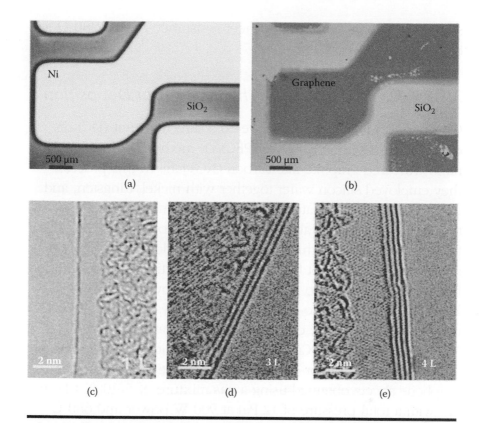

Figure 2.10 (See color insert.) Chemical vapor deposition of graphene on transition metal substrates. Optical microscopic images of (a) the nickel catalyst and (b) the resulting graphene film. TEM images show the nucleation of (c) one, (d) three, or (e) four layers during the growth process. (Reproduced with permission from A. Reina, X.T. Jia, J. Ho, D. Nezich, H.B. Son, V. Bulovic, M.S. Dresselhaus, J. Kong, *Nano Lett.* 9, 30–35, 2009. Copyright 2009 American Chemical Society.)

anywhere between one and three layers and are polycrystalline. Field effect devices are fabricated with epitaxial and CVD graphene display carrier mobilities in excess of 1000 cm²/Vs [40,46]. In the case of CVD graphene, etching of the underlying metal allows the carbon films to be transferred to other substrates. In addition to large-area depositions, graphene grown by the CVD technique has great promise for transparent conducting applications. One such film grown by CVD and transferred via a polydimethylsiloxane (PDMS) stamp onto glass exhibited a sheet resistance of just 280 Ω at 80% optical transmittance [46,47].

2.3.3.3 Plasma-Enhanced Chemical Vapor Deposition

Obraztsov and coworkers discovered a direct current (DC) discharge plasma-enhanced CVD (PECVD) method to produce the nanostructured graphite-like carbon (NG) [48]. In this process, they employed silicon wafer together with nickel, tungsten, and molybdenum as substrates and a gas mixture of CH_4 and H_2 (0% to 25% CH_4), with a total gas pressure of 10 to 150 torr. The NG film thus obtained by this process looks predominantly thicker at several places except for some twisted portions. However, the first report on single- to few-layer graphene by PECVD was during 2004 [49,50]. A radio-frequency PECVD system was employed to synthesize graphene on a variety of substrates (e.g., Si, W, Mo, Zr, Ti, Hf, Nb, Ta, Cr, SiO_2, and Al_2O_3) without any special surface requirements or other catalyst deposition. Further, the graphene sheets obtained using a gas mixture of 5–100% CH_4 in H_2 (with a total pressure of 12 Pa) at 900 W power and 680°C substrate temperature was found to have subnanometer thickness and was erected from the substrate surface. Researchers were attracted to this technique due to its simplicity [49–53].

Zhu et al. proposed a growth mechanism for graphene in a PECVD chamber [52]. Based on their scheme, atomically thin graphene sheets are synthesized by a balance between deposition through surface diffusion of C-bearing growth species from precursor gas and the etching caused by atomic hydrogen. The graphene sheets obtained in this method are vertical due to the plasma electric field direction. However, in another process, Zhang and coworkers showed the synthesis of multilayer graphene nanoflake films (MGNFs) on silicon substrates through microwave PECVD (MW-PECVD) [54]. This method was reported to be a rapid process for growing graphene at a rate of 1.6 μm/min (10 times faster than the other known process). In addition, graphene obtained by this method possessed a highly graphitized knife-edge structure with 2- to 3-nm thickness at the sharp edges; it was roughly vertical to the substrate (Si) and exhibited excellent biosensing capability toward dopamine detection.

Similarly, Yuan et al. synthesized high-quality graphene sheets consisting of one to three layers on a stainless steel substrate at 500°C by MW-PECVD [55]. The process employed a gas mixture of CH_4 and H_2 (1:9 ratio, at a total pressure of 30 torr and a flow rate of 200 sccm) and microwave power of 1200 W. Graphene obtained in this method exhibited better crystallinity than by any other method. The PECVD method has shown the versatility of synthesizing graphene on any substrate, thereby expanding its field of applications. Future developments of this method will definitely bring better control over the thickness of the graphene layers and the scalability.

2.3.4 *Thermal Decomposition*

Thermal decomposition of silicon on the (0001) surface plane of a single crystal of 6H-SiC is also a widely preferred technique to grow epitaxial graphene [56]. Graphene sheets were formed when the H_2-etched surface of 6H-SiC was heated to 1450°C for about 20 min. Graphene usually grown on this surface typically consists of one to three graphene layers, with the number of layers dependent on the decomposition temperature. In a similar process, Rollings and coworkers produced graphene films as thin as the thickness of one atom [42]. Furthermore, the continuous success of this process has attracted the attention of several semiconductor industries (as this process may be a viable technique in the post-CMOS era) [7,39,42,57,58].

Hass and coworkers have already reviewed this topic, dealing with the issues of graphene growth on different faces of SiC and their electronic properties [39]. As a result of rapid progress in this technology, continuous films (millimeter scale) of graphene were synthesized on a nickel thin film coated SiC substrate even at a temperature around 750°C [57, 60]. The advantage of this method is the continuous growth of graphene film over the entire nickel-coated surface. Large-area production of graphene makes this route favorable for industrial application. In a similar process, Emtsev et al. have shown synthesis of large-size, monolayer graphene films even at atmospheric pressure [67]. Further, the method was predicted to yield wafer-size graphene films.

The process of growing graphene on SiC looks attractive, especially for semiconductor industries. However, the shortcomings include controlling the thickness of the graphene layers; repeated production of large-area graphene requires attention prior to adopting the process at an industrial scale. On analyzing the available research works on epitaxial growth of graphene film on an SiC surface, there are a few important issues. For example, graphene grown on SiC(0001) and SiC(0001) was found to have different structures. Unusual rotational stacking was observed in multilayer graphene (up to 60 layers thick) grown on the SiC(0001). The foretold unusual behavior could not be observed in graphene grown on SiC(0001). Future research on this issue should be directed to understanding both the mechanisms of the growth processes and applying the knowledge in developing novel practical devices.

The other important issue is the structure and electronic properties of the interfacial layer between graphene and substrate (as this layer is known to affect the properties of graphene). To date, the effect of the interface has not been well understood; future research should be directed toward understanding this issue. A thorough understanding of the growth mechanism, the interface effects, and the ability to effectively control the number of layers will definitely help in the production of wafer-scale graphene at the bulk industrial scale.

2.3.5 *Thermal Decomposition on Other Substrates*

Other than silicon, thermal decomposition is also possible on other materials. For example, graphene monolayers were grown on single-crystal ruthenium (0001) surface at an ultrahigh-vacuum condition (4×10^{-11} torr) [61]. Prior to this growth procedure, ruthenium crystal was cleaned by repeated cycles of Ar^+ sputtering/annealing followed by exposure to oxygen and heating to high temperature. Soon after this procedure, graphene was found to be formed on the crystal surface, either by thermal decomposition of ethylene (preadsorbed on the crystal surface at room temperature) at 1000 K or by controlled segregation of carbon from the substrate bulk. The single-layer graphene was of high purity, covered a large space (more than several microns), and periodically rippled. In another work, Sutter et al. [47] reported the formation of macroscopic single-crystalline domains of single- to few-layer graphene (more than 200 μm). Furthermore, this is a widely preferred method for the synthesis of graphene on other transition metal surfaces, such as nickel, platinum, cobalt, and so on [62].

2.3.6 Unzipping Multiwall Carbon Nanotubes

Unzipping of MWCNTs can be performed by the intercalation of lithium and ammonia, followed by exfoliation in acid and abrupt heating [63] (see Figure 2.11). The product that results will consist of a mixture of partially open MWNTs and graphene flakes. Unzipping of MWCNTs can also be performed by the plasma etching of MWCNTs partially embedded in a polymer film [64]. The etching procedure will basically open the MWNTs to form graphene. In a different approach, MWNTs were unzipped by a multistep chemical treatment, including exfoliation by concentrated H_2SO_4, $KMnO_4$, and H_2O_2, stepwise oxidation using $KMnO_4$, followed by reduction in NH_4OH and hydrazine monohydrate ($N_2H_4.H_2O$) solution [65]. This new route of unzipping MWCNTs to produce graphene opens the possibilities of synthesizing graphene in a substrate-free manner.

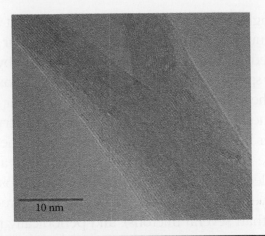

Figure 2.11   TEM image of partially unzipped MWCNT structure. (Reproduced with permission from A.G. Cano-Marquez, F.J. Rodrıguez-Macıas, J. Campos-Delgado, C.G. Espinosa-Gonzalez, F. Tristan-Lopez, D. Ramire-Gonzalez, D.A. Cullen, D.J. Smith, M. Terrones, Y.I. Vega-Cantu, *Nano Lett.* 9, 1527–1533, 2009. Copyright 2009 American Chemical Society.)

2.3.7 *Electrochemical Synthesis*

The electrochemical method is capable of modifying the electronic states via adjusting the external power source to alter the Fermi energy level in the surface of electrode materials. For the first time, Guo et al. [66] reported a facile and fast approach to synthesize high-quality graphene nanosheets in bulk by the electrochemical reduction of exfoliated GO at a graphite electrode. Furthermore, the reaction rate can be accelerated by increasing the reduction temperature, which will also eliminate defects. This electrochemical method has three important advantages: It is (1) fast, (2) greener, and (3) no toxic solvents are used, therefore contamination of the product will not occur. Moreover, the high negative potential can overcome the energy barriers for the reduction of oxygen functionalities (–OH, C–O–C on the plane and –COOH on the edge), thereby effectively reducing the exfoliated GO. The modified electrode can be used in biosensing, fuel cells, and electrocatalysis applications.

Guo et al. [66] employed GO obtained by the Hummers method [20] as the starting material, which consisted of numerous oxygen-bearing groups bonded onto the surface of GO. These oxygen-containing moieties increased the charged capacity, thereby increasing the dispersion of GO in water. Following this, electrochemical reduction of the exfoliated GO was performed on a graphite working electrode at different cathodic potential in a GO dispersion with constant magnetic stirring. Here, the role of magnetic stirring was to propel the GO to the electrode surface and to avoid bubble evolution. The electrochemical setup as well as the optical images of the graphite electrode and the GO suspension before and after electrochemical reduction are illustrated in Figure 2.12. The cyclic voltammograms (Figure 2.13) of a GO-modified glassy carbon electrode (GCE) in a potential window from 0.0 to −1.5 V showed a large cathodic current

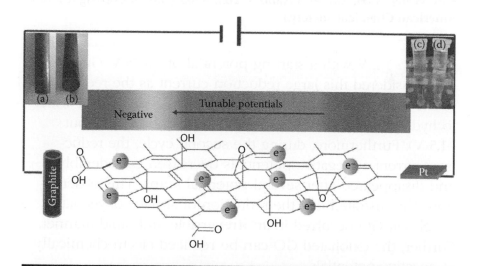

Figure 2.12 Experimental setup and the optical images of the graphite electrode and the GO suspension before (a, c) and after (b, d) electrochemical reduction. The electrochemical reduction potential of the exfoliated GO dispersion at a graphite electrode was −1.5 V versus SCE. (Reproduced with permission from H.-L. Guo, X.-F. Wang, Q.-Y. Qian, F.-B. Wang, X.-H. Xia, *ACS Nano* 3, 2653–2659, 2009. Copyright 2009 American Chemical Society.)

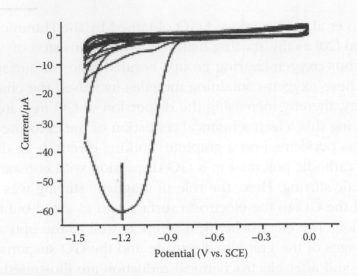

Figure 2.13 Cyclic voltammograms of a GO-modified GCE in PBS (pH 5.0) saturated with nitrogen gas at a scan rate of 50 mV/s. (Reproduced with permission from H.-L. Guo, X.-F. Wang, Q.-Y. Qian, F.-B. Wang, X.-H. Xia, *ACS Nano* 3, 2653–2659, 2009. Copyright 2009 American Chemical Society.)

peak at –1.2 V with a starting potential of –0.75 V. Guo et al. [66] considered this large reduction current as the reduction of the surface oxygen groups because the reduction of water to hydrogen occurs at more negative potentials (at about –1.5 V). Furthermore, during the second cycle, the reduction current at negative potentials decreased considerably and disappeared after several potential scans. This observation demonstrated that the reduction of surface-oxygenated species at GO occurred in an irreversible and rapid manner. Further, the exfoliated GO can be reduced electrochemically at negative potentials.

Guo et al. [66] also performed differential pulse voltammetry (DPV) and cyclic voltammetry (CV) to ascertain the time essential for the electrochemical reduction of exfoliated GO at –1.3 V (vs. SCE) in pH 5.0 phosphate-buffered saline (PBS). Based on this experiment, Guo's group [66] demonstrated that although at –1.3 V (vs. Saturated Calomel Electrode (SCE)) GO can be reduced electrochemically, the process is slow. So, they performed several trial sets and finally found that when the potential was set at −1.5 V (vs. SCE), it resulted in the attachment of a bulk quantity of black precipitate onto the bare graphite electrode due to the reduction of oxygenated GO. Further, the yellow color of the native GO suspension turned colorless.

2.3.8 Other Available Methods

Although the methods discussed are routinely employed for graphene synthesis, there are also a few other techniques in practice to synthesize graphene. Kim et al. [67] employed aluminum sulfide (Al_2S_3) and calcined it under a (CO + Ar) gaseous environment. Under these conditions, CO was reduced by Al_2S_3 to form gaseous carbon and α-alumina. Finally, multilayer graphene sheets were crystallized on the alumina particles. However, the mechanism for this transformation is not yet well understood or predicted; the process seems to be interesting due to its simplicity. In the future, proper tuning of various parameters will efficiently minimize and control the number of graphene layers.

In another work, Zhang et al. [10] demonstrated the possibility of cleaving HOPG using a microcantilever (such as an AFM tip) to obtain thin graphitic sheets. However, the product was 10- to 100-nm thick and hence could not be treated as graphene. Nevertheless, the effort has shown a new possibility of producing graphene sheets. In another recent approach, conducting nanocarbon films (thickness about 1 nm) were produced through a complex processing route based on molecular self-assembly, electron irradiation, and pyrolysis [68,69].

At this stage, these processes are not easy to adopt industrially. However, in the future, with rapid progress in graphene research these techniques will be made easier to follow and to synthesize graphene in bulk quantities.

References

1. V. Eswaraiah, S. S. J. Aravind, S. Ramaprabhu, Top down method of synthesis of highly conducting graphene by exfoliation of graphite oxide using focused solar radiation, *J. Mater. Chem.* 21, 6800–6803, 2005.
2. J.H. Lee, D.W. Shin, V.G. Makotchenko, A.S. Nazarov, V.E. Fedorov, Y.H. Kim, J.Y. Choi, J.M. Kim, J.-B. Yoo, One-step exfoliation synthesis of easily soluble graphite and transparent conducting graphene sheets, *Adv. Mater.* 21, 4383–4387 (2009).
3. K.S. Novoselov, D. Jiang, F. Schedin, T.J. Booth, V.V. Khotkevich, S.V. Morozov, A.K. Geim, Two dimensional atomic crystals, *Proc. Natl. Acad. Sci. USA* 102, 10451–10453 (2005).
4. J.C. Meyer, A.K. Geim, M.I. Katsnelson, K.S. Novoselov, T.J. Booth, S. Roth, The structure of suspended graphene sheets, *Nature* 446, 60–63 (2007).
5. S. Stankovich, R.D. Piner, X.Q. Chen, N.Q. Wu, S.B.T. Nguyen, R.S. Ruoff, Stable aqueous dispersions of graphitic nanoplatelets *via* the reduction of exfoliated graphite oxide in the presence of poly(sodium 4-styrenesulfonate), *J. Mater. Chem.* 16, 155–158 (2006).
6. C. Berger, Z.M. Song, X.B. Li, X.S. Wu, N. Brown, C. Naud, D. Mayou, T.B. Li, J. Hass, A.N. Marchenkov, E.H. Conrad, P.N. First, W.A. De Heer, Electronic confinement and coherence in patterned epitaxial graphene, *Science* 312, 1191 (2006).
7. C. Berger, Z.M. Song, T.B. Li, X.B. Li, A.Y. Ogbazghi, R. Feng, Z.T. Dai, A.N. Marchenkov, E.H. Conrad, P.N. First, W.A. De Heer, Ultrathin epitaxial graphite: 2D electron gas properties and a route toward graphene-based nanoelectronics, *J. Phys. Chem. B* 108, 19912–19916.
8. G. Eda, G. Fanchini, M. Chhowalla, Large-area ultrathin films of reduced graphene oxide as a transparent and flexible electronic material, *Nat. Nanotechol.* 3, 270–274 (2008).
9. X.K. Lu, M.F. Yu, H. Huang, R.S. Ruoff, Tailoring graphite with the goal of achieving single sheets, *Nanotechnology* 10, 269 (1999).
10. Y.B. Zhang, J.P. Small, W.V. Pontius, P. Kim, Fabrication and electric-field-dependent transport measurements of mesoscopic graphite devices, *Appl. Phys. Lett.* 86, 073104 (2005).

11. A.M. Affoune, B.L.V. Prasad, H. Sato, T. Enoki, Y. Kaburagi, Y. Hishiyama, Experimental evidence of a single nano-graphene, *Chem. Phys. Lett.* 348, 17–20 (2001).

12. M.J. Allen, V.C. Tung, R.B. Kaner, Honey comb graphene: A review of graphene, *Chem. Rev.* 110, 132 (2010).

13. R.R. Nair, P. Blake, A.N. Grigorenko, K.S. Novoselov, T.J. Booth, T. Stauber, N.M.R. Peres, A.K. Geim, Fine structure constant defines visual transparency of graphene, Science 320, 1308 (2008).

14. T. Stauber, N.M.R. Peres, A.K. Geim, Optical conductivity of graphene in the visible region of the spectrum, *Phys. Rev. B* 78, 085432 (2008).

15. P. Blake, E.W. Hill, A.H.C. Neto, K.S. Novoselov, D. Jiang, R. Yang, T.J. Booth, A.K. Geim, Making graphene visible, *Appl. Phys. Lett* 91, 063124 (2007).

16. S. Stankovich, D.A. Dikin, R.D. Piner, K.A. Kohlhaas, A. Kleinhammes, Y. Jia, Y. Wu, S.T. Nguyen, R.S. Ruoff, Synthesis of graphene-based nanosheets *via* chemical reduction of exfoliated graphite oxide, *Carbon* 45, 1558–1565 (2007).

17. I. Jung, D.A. Dikin, R.D. Piner, R.S. Ruoff, Tunable electrical conductivity of individual graphene oxide sheets reduced at "low" temperatures, *Nano Lett.* 8, 4283–4287 (2008).

18. D. Yang, A. Velamakanni, G. Bozoklu, S. Park, M. Stoller, R.D. Piner, S. Stankovich, I. Jung, D.A. Field, C.A. Ventrice, R.S. Ruoff, Chemical analysis of graphene oxide films after heat and chemical treatments by X-ray photoelectron and micro-Raman spectroscopy, *Carbon* 47, 145–152 (2009).

19. H.-K. Jeong, Y.P. Lee, R.J.W.E. Lahaye, M.-H. Park, K.H. An, I.J. Kim, C.-W. Yang, C.Y. Park, R.S. Ruoff, Y.H. Lee, Evidence of graphitic AB stacking order of graphite oxides, *J. Am. Chem. Soc.* 130, 1362–1366 (2008).

20. W.S. Hummers, R.E. Offeman, Preparation of graphitic oxide, *J. Am. Chem. Soc.* 80, 1339 (1958).

21. V.C. Tung, M.J. Allen, Y. Yang, R.B. Kaner, High-throughput solution processing of large-scale graphene, *Nat. Nanotechnol.* 4, 25–29 (2009).

22. M.J. Allen, J.D. Fowler, V.C. Tung, Y. Yang, B.H. Weiller, R.B. Kaner, Temperature dependent Raman spectroscopy of chemically derived graphene, *Appl. Phys. Lett.* 93, 193119 (2008).

23. J.D. Fowler, M.J. Allen, V.C. Tung, Y. Yang, R.B. Kaner, B.H. Weiller, Practical chemical censors from chemically derived graphene, *ACS Nano* 3, 301–306 (2009).

24. E.W. Schmidt, *Hydrazine and its derivatives*, Wiley-Interscience, New York (2001).

25. D.A. Dikin, S. Stankovich, E.J. Zimney, R.D. Piner, G.H.B. Dommett, G. Evmenenko, S.T. Nguyen, R.S. Ruoff, Preparation and characterization of graphene oxide paper, *Nature* 448, 457–460 (2007).

26. D. Li, R.B. Kaner, Material science: Graphene based materials, *Science* 320, 1170–1171 (2008).

27. S. Park, K.S. Lee, G. Bozoklu, W. Cai, S.T. Nguyen, R.S. Ruoff, Graphene oxide papers modified by divalent ions-enhancing mechanical properties via chemical cross-linking, *ACS Nano* 2, 572–578 (2008).

28. N. Tyutyulkov, G. Madjarova, F. Dietz, K. Mullen, Is 2-D graphite an ultimate large hydrocarbon? 1. Energy spectra of giant polycyclic aromatic hydrocarbons, *J. Phys. Chem. B* 102, 10183–10189 (1998).

29. X.Y. Yang, X. Dou, A. Rouhanipour, L.J. Zhi, H.J. Rader, K.J. Mullen, Two dimensional graphene nanoribbons, *J. Am. Chem. Soc.* 130, 4216 (2008).

30. A.J. Berresheim, M. Muller, K. Mullen, Polyphenylene nano-structures, *Chem. Rev.* 99, 1747–1785 (1999).

31. F. Dotz, J.D. Brand, S. Ito, L. Gherghel, K. Mullen, Synthesis of large polycyclic aromatic hydrocarbons: Variation of size and periphery, *J. Am. Chem. Soc.* 122, 7707–7717 (2000).

32. M.D. Watson, A. Fechtenkotter, K. Mullen, Big is beautiful—Aromaticity revisited from the view point of macromolecular and supramolecular benzene chemistry, *Chem. Rev.* 101, 1267–1300 (2001).

33. I. Gutman, Z. Tomovic, K. Mullen, E.P. Rabe, On the distribution of π-electrons in large polycyclic aromatic hydrocarbons, *Chem. Phys. Lett.* 397, 412–416 (2004).

34. J.S. Wu, W. Pisula, K. Mullen, Graphene as a potential material for electronics, *Chem. Rev.* 107, 718–747 (2007).

35. L.J. Cote, F. Kim, J. Huang, Langmuir-Blodgett assembly of graphitic oxide single layers, *J. Am. Chem. Soc.* 131, 1043–1049 (2009).

36. J.H. Wu, Q.W. Tang, H. Sun, J.M. Lin, H.Y. Ao, M.L. Huang, Y.F. Huang, Conducting film from graphite oxide nanoplatelets and poly(acrylic acid) by layer-by-layer self-assembly, *Langmuir* 24, 4800–4805 (2008).

37. X. Li, G. Zhang, X. Bai, X. Sun, X. Wang, E. Wang, H. Dai, Highly conducting graphene films and Langmuir-Blodgett film, *Nat. Nanotechnol.* 3, 538–542 (2008).

38. W.A. De Heer, C. Berger, X.S. Wu, P.N. First, E.H. Conrad, X.B. Li, T.B. Li, M. Sprinkle, J. Hass, M.L. Sadowski, M. Potemski, G. Martinez, Epitaxial graphene, *Solid State Commun.* 143, 92–100 (2007).

39. E. Rollings, G.H. Gweon, S.Y. Zhou, B.S. Mun, J.L. McChesney, B.S. Hussain, A.V. Fedorov, P.N. First, W.A. De Heer, A. Lanzara, Synthesis and characterization of atomically thin graphite films on a silicon carbide substrate, *J. Phys. Chem. Solids* 67, 2172–2177 (2006).

40. J. Kedzierski, P.L. Hsu, P. Healey, P.W. Wyatt, C.L. Keast, M. Sprinkle, C. Berger, W.A. De Heer, Epitaxial graphene substrates on SiC substrates, *IEEE Trans. Electron Devices* 55, 2078–2085 (2008).

41. C. Berger, Z.M. Song, X.B. Li, X.S. Wu, N. Brown, D. Maud, C. Naud, W.A. Heer, Magnetotransport in high mobility epitaxial graphene, *Phys. Status Solidi A: Appl. Mater. Sci.* 204, 1746–1750 (2007).

42. E. Rollings, G.H. Gweon, S.Y. Zhou, B.S. Mun, J.L. McChesney, B.S. Hussain, A.V. Fedorov, P.N. First, W.A. De Heer, A. Lanzara, Synthesis and characterization of atomically thin graphite films on silicon carbide substrate, *J. Phys. Chem. Solids* 67, 2172–2177 (2006).

43. V.W. Brar, Y. Zhang, Y. Yayon, T. Ohta, J.L. McChesney, A. Bostwick, E. Rotenberg, K. Horn, M.F. Crommie, Scanning tunneling spectroscopy of in-homogenous electronic structure in monolayer and bilayer graphene on SiC, *Appl. Phys. Lett.* 91, 122102 (2007).

44. E. Rotenberg, A. Bostwick, T. Ohta, J.L. McChesney, T. Seyller, K. Horn, Origin of the energy bandgap in epitaxial graphene, *Nat. Mater.* 7, 258–259 (2008).

45. A. Reina, X.T. Jia, J. Ho, D. Nezich, H.B. Son, V. Bulovic, M.S. Dresselhaus, J. Kong, Large area few-layer graphene films on arbitrary substrates by chemical vapor deposition, *Nano Lett.* 9, 30–35 (2009).

46. K.S. Kim, Y. Zhao, H. Jang, S.Y. Lee, J.M. Kim, K.S. Kim, J.H. Ahn, P. Kim, J.Y. Choi, J.B.H. Hong, Large-scale pattern growth of graphene films for stretchable transparent electrodes, *Nature* 457, 706–710 (2009).

47. P.W. Sutter, J.I. Flege, E.A. Sutter, Epitaxial graphene on ruthenium, *Nat. Mater.* 7, 406–411 (2008).

48. A.N. Obraztsov, A.A. Zolotukhin, A.O. Ustinov, A.P. Volkov, Y. Svirko, K. Jefimovs, DC discharge plasma studies for nano-structured carbon CVD, *Diamond Related Mater.* 12, 917–920 (2003).

49. J.J. Wang, M.Y. Zhu, R.A. Outlaw, X. Zhao, D.M. Manos, B.C. Holoway, Free-standing subnanometer graphite sheets, *Appl. Phys. Lett.* 85, 1265 (2004).

50. J.J. Wang, M.Y. Zhu, R.A. Outlaw, X. Zhao, D.M. Manos, B.C. Holoway, Synthesis of carbon nanosheets by inductively coupled radio-frequency plasma enhanced chemical vapor deposition, *Carbon* 42, 2867–2872 (2004).

51. M. Hiramatsu, K. Shiji, H. Amano, M. Hori, Fabrication of vertically aligned carbon nanowalls using capacitively coupled plasma-enhanced chemical vapor deposition assisted by hydrogen radical injection, *Appl. Phys. Lett.* 84, 4708–4710 (2004).

52. M. Zhu, J. Wang, B.C. Holloway, R.A. Outlaw, X. Zhao, K. Hou, V. Shutthanandan, D.M. Manos, A mechanism for carbon nanosheet formation, *Carbon* 45, 2229–2234 (2007).

53. C. Wang, S. Yang, Q. Wang, Z. Wang, J. Zhang, Super-low friction and super-elastic hydrogenated carbon films originated from a unique fullerene-like nanostructure, *Nanotechnology* 19, 225709 (2008).

54. A. Malesevic, R. Kemps, L. Zhang, R. Erni, G. Van Tendeloo, A. Vanhulsel, C. Van Haesendonck, A versatile plasma tool for the synthesis of carbon nanotubes and few-layer graphene sheets, *J. Optoelect. Adv. Mater.* 10, 2052–2055 (2008).

55. G.D. Yuan, W.J. Zhang, Y. Yang, Y.B. Tang, Y.Q. Li, J.X. Wang, X.M. Meng, Z.B. He, C.M.L. Wu, I. Bello, C.S. Lee, S.T. Lee, Graphene sheets *via* microwave chemical vapor deposition, *Chem. Phys. Lett.* 467, 361–364 (2009).

56. Z. Wu, W. Ren, L. Gao, B. Liu, C. Jiang, H. Cheng, Synthesis of high-quality graphene with a pre-determined number of layers, *Carbon* 47, 493–499 (2009).

57. T. Ohta, F.E. Gabaly, A. Bostwick, J.L. McChesney, K.V. Emtsev, A.K. Schmid, T. Seyller, K. Horn, E. Rotenberg, Morphology of graphene thin film growth on SiC(0001), *New J. Phys.* 10, 023034 (2008).

58. Z.G. Cambaz, G. Yushin, S. Osswald, V. Mochalin, Y. Gogotsi, Noncatalytic synthesis of carbon nanotubes graphene and graphite on SiC, *Carbon* 46, 841–849 (2008).

59. Z.-Y. Juang, C.-Y. Wu, C.-W. Lo, W.-Y. Chen, C.-F. Huang, J.-C. Hwang, F.-R. Chen, K.C. Leou, C.H. Tsai, Synthesis of graphene on silicon carbide substrates at low temperature, *Carbon* 47, 2026–2031 (2009).

60. K.V. Emtsev, A. Bostwick, K. Horn, J. Jobst, G.L. Kellogg, l. Ley, J.L. McChesney, T. Ohta, S.A. Reshanov, J. Rohr, E. Rotenberg, A.K. Schmid, D. Waidmann, H.B. Webber, T. Seyller, Towards wafer-size graphene layers by atmospheric pressure graphitization of silicon carbide, *Nat. Mater.* 8, 203–207 (2009).

61. A.L. V´azquez de Parga, F. Calleja, B. Borca, M.C.G. Passeggi Jr., J.J. Hinarejos, F. Guinea, R. Miranda, Periodically rippled graphene: growth and spatially resolved electronic structure, *Phys. Rev. Lett.* 100, 056807 (2008).

62. J. Wintterlin, M.-L. Bocquet, Graphene on metal surfaces, *Surf. Sci.* 603, 1841 (2009).

63. A.G. Cano-Marquez, F.J. Rodriguez-Macias, J. Campos-Delgado, C.G. Espinosa-Gonzalez, F. Tristan-Lopez, D. Ramire-Gonzalez, D.A. Cullen, D.J. Smith, M. Terrones, Y.I. Vega-Cantu, Ex-MWNTs: Graphene sheets and ribbons produced by lithium intercalation and exfoliation of carbon nanotubes, *Nano Lett.* 9, 1527–1533 (2009).

64. L. Jiao, L. Zhang, X. Wang, G. Diankov, H. Dai, Narrow graphene nanoribbons from carbon nanotubes, *Nature* 458, 877–880 (2009).

65. D.V. Kosynkin, A.L. Higginbotham, A. Sinitskii, J.R. Lomeda, A. Dimiev, B.K. Price, J.M. Tour, Longitudinal unzipping of carbon nanotubes to form graphene nanoribbons, *Nature* 458, 872–876 (2009).

66. H.-L. Guo, X.-F. Wang, Q.-Y. Qian, F.-B. Wang, X.-H. Xia, A green approach to the synthesis of graphene nanosheets, *ACS Nano* 3, 2653–2659 (2009).

67. C.D. Kim, B.K. Min, W.S. Jung, Preparation of graphene sheets by the reduction of carbon monoxide, *Carbon* 47, 1610–1612 (2009).

68. A. Turchanin, A. Beyer, C.T. Nottbohm, X. Zhang, R. Stosch, A. Sologubenko, J. Mayer, P. Hinze, T. Weimann, A. Golzhauser, One nanometer thin carbon nanosheets with tunable conductivity and stiffness, *Adv. Mater.* 21, 1233–1237 (2009).

69. M.J. Schultz, X. Zhang, S. Unarunotai, D.-Y. Khang, Q. Cao, C. Wang, C. Lei, S. MacLaren, J.A.N.T. Soares, I. Petrov, J.S. Moore, J.A. Rogers, Synthesis of linked carbon monolayers: Films, balloons, tubes and pleated sheets, *Proc. Natl. Acad. Sci. USA* 105, 7353–7358 (2008).

Chapter 3

Surface Characterization of Graphene

3.1 Graphene Characterization

The characterization of graphene can be performed by a variety of techniques, such as scanning electron microscopy (SEM), scanning tunneling microscopy (STM), X-ray photoelectron spectroscopy (XPS), atomic force microscopy (AFM), Raman spectroscopy, and X-ray diffraction (XRD). However, the characterization is not limited to the techniques mentioned. This chapter discusses briefly certain important characterization techniques often employed in most laboratories.

3.1.1 Optical Imaging of Graphene Layers

Single-layer, bilayer, and few-layer graphene can be imaged using techniques such as optical microscopy, AFM, SEM, and high-resolution transmission electron microscopy (HRTEM) [1–4]. For better presentation of the different layers of graphene, a combination of two or more imaging techniques is often employed [1–4]. Optical microscopy is widely employed to image the different layers of graphene because it is the cheapest, nondestructive, and readily available method in the laboratories. To perform optical microscopy, graphene layers should be mounted on silicon dioxide substrate for excellent contrast imaging [1,4]. Since 2008, substantial attention has been given to design substrates to enhance the visibility of thin sheets [1–4]. Mechanisms associated with such contrast can be explained in terms of Fabry-Perot interference in the dielectric surface layer, which governs the fluorescence intensity, thereby allowing contrast between graphene layers and substrate [1].

SiO_2 and Si_3N_4 are widely preferred overlay materials on silicon to enhance the contrast of dielectric graphene layers [1,4,5]. The other factor associated with the contrast and that modulates the contrast is the wavelength of the incident light [1,4]. Blake et al. [5] demonstrated the contrast variation with the help of different narrowband filters to detect sheets for any thickness of SiO_2 support [1,4]. Further, they witnessed that under the illumination of normal white light, graphene sheets were invisible on 200-nm SiO_2 [4,5]. However, thin sheets became visible on 300-nm SiO_2 on illumination with green light, and thick sheets were visible on 200-nm SiO_2 on illumination using a blue light [1,4].

Figure 3.1 is the optical image of the different layers of the micromechanically exfoliated graphene on silicon substrate with a 300-nm SiO_2 overlayer. A color contrast method and AFM techniques revealed the number of layers in the graphene sample [1,4,6]. Further, the detection technique demonstrated was dependent on the thickness of the substrate and the wavelength of the incident light [1]. Intensive additional research is essential to facilitate the visualization of graphene-based sheets that are independent of support material [1].

3.1.2 Fluorescence Quenching Technique

Graphene, RGO (reduced graphene oxide), and GO (graphene oxide) can be imaged by a fluorescence quenching microscopic (FQM) technique. Imaging the samples by FQM will be helpful for immediate sample evaluation and manipulation

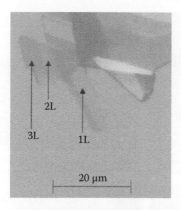

Figure 3.1 Optical microscopic image of single-, double-, and triple-layer graphene on silicon with a 300-nm SiO$_2$ overlayer, labeled as 1L, 2L, and 3L, respectively. (Reproduced with permission from J.S. Park, A. Reina, R. Saito, J. Kong, G. Dresselhaus, M.S. Dresselhaus, *Carbon* **47, 1303–1310, 2009.)**

to improve the synthesis process [1,4,7]. FQM has been proved to be a low-cost, time-saving method to visualize GO and RGO [1,4]. The imaging mechanism involves quenching the emission from dye-coated GO and RGO [1,4]. Later, the dye can be removed by rinsing without disrupting the sheets. The appearance of the contrast is attributed to the chemical interaction between the GO and the dye molecule [1]. The charge transfer arising from dye molecule to GO causes quenching of fluorescence [1,4,8]. The contrast measured in the fluorescence images (Figure 3.1) was achieved for a 300-nm SiO$_x$ layer. An increase in the contrast was noticed when GO sheets were deposited on 100-nm SiO$_x$. Contrast values of 0.52 and 0.07 were noticed for the quartz and glass, respectively [1,4]. These values were sufficient for clear visualization [1,4,8].

Figure 3.2 shows an image obtained from FQM; it is compared with the AFM image of the same area. The technique has the capability of visualizing the microstructures of the GO/RGO film even on plastic substrate [1]. This technique is based on light, so it is restricted to samples that contain layers with a micrometer length [1]. Further, the lateral resolution

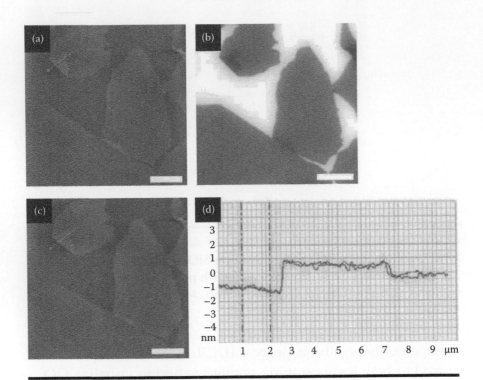

Figure 3.2 (a) AFM image showing GO single layers deposited on a SiO₂/Si wafer before applying a 30-nm thick fluorescein/polyvinyl-pyrrolidone layer for FQM. (b) A FQM image of the same area of the wafer showing good correlation to the AFM view. (c) After washing off the dye coating, no residues can be detected by AFM. (d) Line scan data on a folded sheet show no significant deviation in thickness before and after FQM imaging (all scale bars 10 μm). (Reproduced with permission from J. Kim, L.J. Cote, F. Kim, J. Huang, *J. Am. Chem. Soc.* 132, 260–267, 2009.)

of FQM is diffraction limited. FQM necessitates the addition of dye to the graphene surface. The disadvantage of this dye addition is that it prohibits the further use of the same sample due to the attachment of unwanted functional groups [1,4].

3.1.3 Atomic Force Microscopy

AFM is an extremely useful technique to probe the thickness of layers at the nanometer scale [1]. However, this technique will be cumbersome for imaging large-area graphene. Moreover, AFM imaging gives only topographic contrast, which cannot distinguish between the graphene oxide and the graphene layers in normal operation [1]. Moreover, phase imaging is one of the attractive features of tapping-mode AFM; this facilitates distinguishing between defect-free pristine graphene and its functionalized version [1]. One of the reasons is the difference in the interaction forces between the AFM tip and the attached functional group.

Paredes et al. [9] demonstrated the influence of the attractive mode of AFM to determine the thickness of the sheet. Further, Paredes et al. [9] found that the repulsive mode induced deformation, causing error in the height measurement, and observed a thickness of 1.0 nm for the unreduced graphene oxide and 0.6 nm for the chemically reduced GO. The difference in the thickness and the phase contrast was reported to be caused by a hydrophilicity difference resulting from a distinct oxygen functional group in the reduction process, as shown in Figure 3.3.

Besides imaging and thickness detection, AFM has been explored for mechanical characterization of graphene as it can resolve the small forces involved in the deformation process [1]. A variety of AFM modes allowed the study of mechanical, frictional, electrical, magnetic, and even elastic properties of graphene flakes [1,10].

3.1.4 Transmission Electron Microscopy

In general, transmission electron microscopy TEM is frequently used to image nanosize materials to atomic-scale resolution. In TEM, a transmitted electron beam passes through an ultrathin sample and reaches the imaging lenses and detector [1]. Graphene and RGO contain a layer that is one atom thick [1]. TEM is the only reliable tool that can resolve the atomic features of graphene. However, use of traditional TEM means is limited by their resolution at low operating voltage, whereas operation at high voltage damages the monolayer.

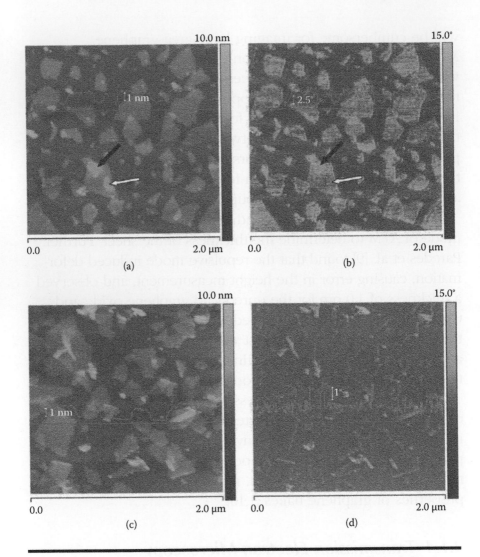

Figure 3.3   Height (a and c) and corresponding phase (b and d) tapping-mode AFM images of unreduced (a and b) and chemically reduced (c and d) graphene oxide nanosheets deposited from aqueous dispersions onto freshly cleaved highly oriented pyrolytic graphite (HOPG). The images were recorded in the attractive regime of tip sample interaction. Superimposed onto each image is a line profile taken along the marked red line. (Reproduced with permission from J.I. Paredes, S. Villar-Rodil, P. Solis-Fernandez, A. Martinez-Alonso, J.M.D. Tascon, *Langmuir* 25, 5957–5968, 2009.)

A few researchers have used a new class of TEM that is aberration corrected in combination with a monochromator; it can provide 1-Å resolution at an acceleration voltage of only 80 kV [1,11,12]. For the first time, Mayer's group has shown direct high-resolution images of the graphene lattice depicting every single carbon atom arranged in hexagonal fashion [1,13]. This clearly reveals the ball-and-stick model, where bright and dark contrasts in the image correspond to the atoms and the gaps, respectively (shown in Figure 3.4). Researchers also indicated that the imperfections and topological peculiarities in graphene affected the electronic and mechanical properties, which can be determined using such an aberration-corrected, low-voltage TEM [1].

In another imaging technique, Gass et al. [14] have shown atomic lattice, defects, and surface contamination in high-angle annular dark-field (HAADF) images using scanning transmission electron microscopy (STEM). The technique involves focusing an electron beam onto a monoatomic region and further scanning [1,14]. This technique easily detects the arrangement of atomic-scale defects and contaminant atoms using Z contrast. Figure 3.5 shows the HRTEM bright-field and HAADF image of a monolayer, revealing a clean graphene monolayer surrounded by the contaminants, and mono- and divacancy defects due to missing carbon atoms [1].

It is also worth mentioning here that due to the transparent nature of the graphene and its oxide, these are used as a support for imaging and dynamics of light atoms and molecules under TEM [1,15–17]. Further, conventional TEM can be utilized to observe the smallest atom and molecule using graphene as a membrane to support other samples in TEM [1]. Because graphene is one atom thick, it provides the thinnest possible continuous support, unlike other amorphous TEM support [1]. Its crystallinity and high conductivity facilitate background subtraction and charging reduction [1].

As mentioned in the previous section, RGO has superior conductivity compared to GO, but it is inferior to the

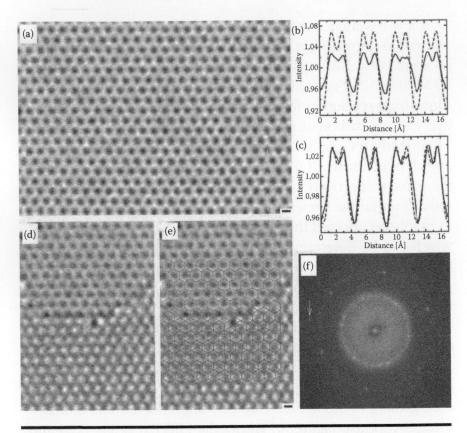

Figure 3.4 (a) Direct image of a single-layer graphene membrane. (b) Contrast profile along the dotted line in panel a (solid) along with a simulated profile (dashed). The experimental contrast is a factor of 2 smaller: Panel (c) shows the same experimental profile with the simulated contrast scaled down by a factor of 2. (d and e) Step from a monolayer (upper part) to a bilayer (lower part of the image) showing the unique appearance of the monolayer. (e) The same image with an overlay of the graphene lattice and the second layer, offset in the Bernal (ab) stacking of graphite. (f) Numerical diffractogram calculated from an image of the bilayer region. The outermost peaks, one of them indicated by the arrow, correspond to a resolution of 1.06 Å. The scale bars are 2 Å. (Reproduced with permission from J.C. Meyer, C. Kisielowski, R. Erni, M.D. Rossell, M.F. Crommie, A. Zettl, *Nano Lett.* 8, 3582–3586, 2008.)

Figure 3.5 High-resolution images of monolayer graphene. Bright-field (a) and HAADF (b) images of the monolayer showing a clean patch of graphene surrounded by a monoatomic surface layer; individual contaminant atoms of higher atomic number can be seen in (b). The inset Fast Fourier Transform (FFT) shows the lattice in the HAADF image and, by applying a band-pass filter, the atomic structure is apparent. HAADF lattice images of defects showing a monovacancy (c) and a divacancy (d). Inset shows the FFT of the raw image. (Reproduced with permission from M.H. Gass, U. Bangert, A.L. Bleloch, P. Wang, R.R. Nair, A.K. Geim, *Nat. Nanotechnol.* 3, 676–681, 2008.)

pristine graphene [1,4]. It is expected that during the reduction process, many defects are generated in the carbon two-dimensional lattice (point defect or incomplete removal of epoxy/oxygenated functional groups) [1]. These defects and foreign functional groups remarkably affect the electronic and thermal conductivity [1]. So, it is important to identify the most dominant defects in graphene before incorporation in technologically important electronic devices [1]. On combination with other spectroscopic techniques, HRTEM will expose the atomic structure of GO and RGO [1]. Recently, the local chemical structure and defect structure of RGO and GO have been studied in more detail at the atomic level [1,11,18].

3.1.5 Raman Spectroscopy

The identity of carbon allotropes can be determined by Raman spectroscopy by the appearance of D, G, and 2D peaks around 1350, 1580, and 2700 cm^{-1}, respectively, due to the change in electron bands [1,4]. Identification of these features allows characterization of graphene layers in terms of number of layers present and their effect of strain, doping concentration, and temperature and the presence of defects [1,4]. The G band is associated to the doubly degenerated E_{2g} phonon mode at the Brillouin zone center. This band (near 1580 cm^{-1}) is due to the in-plane vibration of the sp^2 carbon atoms, whereas the 2D band is at almost double the frequency of the D band and originates from a second-order Raman scattering process [1,4]. The D band appears due to the presence of disorder in atomic arrangement or an edge effect of graphene, ripples, and charge puddles [1].

Comparison of Raman spectra between graphite and single- and few-layer graphene is shown in Figure 3.6 [1,19]. The Raman spectra at the center of the graphene layers do not show a D peak, which confirms the absence of defects. Instead, the significant change in shape and intensity was observed for a 2D band of graphene and graphite.

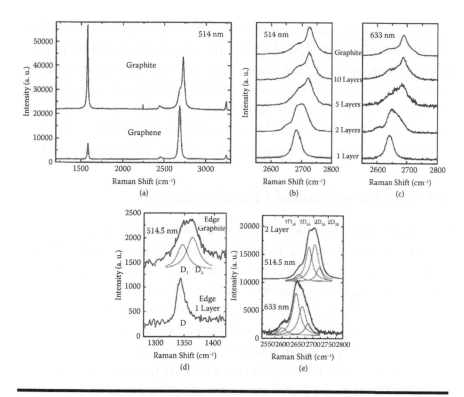

Figure 3.6 **Comparison of Raman spectra at 514 nm for the graphite and single-layer graphene. (b) and (c) Evolution in 2D band as a function of layers at 514- and 633-nm excitations. (d) and (e) Comparison of the D band at 514 nm at the edge of bulk graphite and single-layer graphene. The fit of the D1 and D2 components of the D band of bulk graphite is shown. (e) The four components of the 2D band in two-layer graphene at 514 and 633 nm. (Reproduced with permission from A.C. Ferrari, J.C. Meyer, V. Scardaci, C. Casiraghi, M. Lazzeri, F. Mauri, et al.,** *Phys. Rev. Lett.* **97, 187401, 2006.)**

Ferrari et al. [19] have shown that the 2D band splits into two components for bulk graphite and four components for bilayer graphene (Figures 3.6c and 3.6d). An increase in the number of layers reduces the relative intensity of 2D and increases its FWHM (full width at half maximum) and makes it blue shifted [1,6,19,20]. The single sharp 2D peak was reported for a monolayer that was four times more intense than the G peak [1].

The properties of graphene crucially depend on the number of layers and purity [1,4]. As a result, various researchers have used Raman spectra as a nondestructive tool to characterize and maintain quality control of mono- and few-layer graphene [1]. Various effects on graphene have also been studied by following trends in Raman spectra, including thickness determination, strain in graphene layers, defects, and doping [1,4].

3.1.6 Electrochemical Characterization

Graphene sheets provide a 2D environment for favorable electron transport [2,21]. Heterogeneous electron transfer from or to a graphene sheet occurs at the edges of the graphene sheet; heterogeneous electron transfer from the plane of a graphene sheet is close to zero [2,22]. Oxygen-containing groups present at the graphene edges influence its electrochemistry [2]. Nevertheless, it is uncertain if this influence is positive (i.e., enhances the heterogeneous electron transfer rate) or negative (retards the heterogeneous electron transfer rate) [2].

Chou et al. [23] provided evidence that the modification of carboxylic groups at the electrodes by single-wall carbon nanotubes is responsible for increasing the heterogeneous electron transfer rate constant (and thus for increasing the speed of electron transfer) of ferro-/ferricyanide [2]. However, Pumera contradicted this in his work [24]. In another work, Ji et al. [25] indicated that the heterogeneous electron transfer between the graphitic material and ferro/ferricyanide decreased with an increase in oxygen-containing groups on the graphitic material. On the other hand, Pumera et al. [26] showed that oxygen-containing groups on carbon nanotubes and graphite possess preferential "electrocatalytic" properties for the oxidation of endiol groups (Figure 3.7). In the same work, the authors described that this electron transfer enhancement was due to the presence of oxygen-containing species on the graphene sheets generated during acid treatment [2,26].

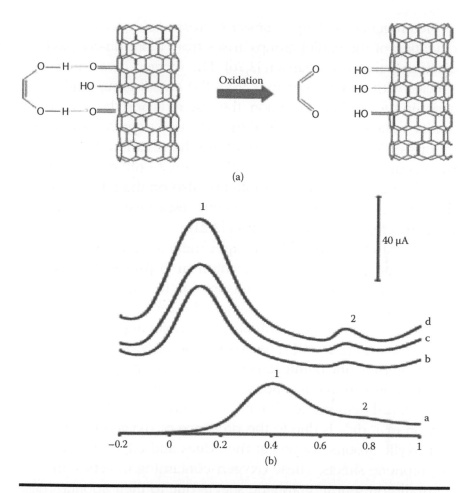

Figure 3.7 **Electrocatalysis toward oxidation of endiol groups. (a) The presence of oxygen-bearing moieties on the surface of graphene sheets offers enhanced electron transfer for the oxidation of endiols via a proton-assisted electron transfer mechanism, thereby confirming their involvement during the surface preparation on electrocatalysis. (b) Differential pulse voltammetry (DPV) in real samples of becozyme containing ascorbic acid (1) and pyridoxine (2), using different screen-printed electrodes (SPEs): (a) bare; (b) SPE activated; (c) SPE graphite; and (d) SPE multiwall carbon nanotube (MWCNT). (Reproduced with permission from A.G. Crevillen, M. Pumera, M.C. Gonzalez, A. Escarpa, *Analyst* 134, 657–662, 2009.)**

Heterogeneous charge transfer corresponding to the oxidation of the endiol groups arises from a proton-coupled electron transfer mechanism [2,26]. The oxygen-containing species withdraw two protons from the endiol group and assist in the oxidation reaction, thereby reducing the overpotential voltage [2,26]. However, the effect of oxygen-containing groups on the electrochemistry of graphene cannot be overestimated [2,21]. It is their influence not only on the heterogeneous electron transfer rate but also on the adsorption/desorption of molecules, which occur respectively before and after the electrochemical reaction [2,21].

Pumera et al. proved both experimentally and theoretically factors that govern the mechanism of β-nicotinamide adenine dinucleotide (NAD+) adsorption to the graphene sheets [2,27]. NAD+ is a key element in electrochemical enzyme biosensors and biofuel cells employing dehydrogenase enzymes [2,28–31]. The adsorption of NAD+ on the carbon materials (including carbon nanotubes and graphite) is a serious concern and has not yet been studied in detail [2,21]. Pumera [21] demonstrated that the adsorption of NAD+ at sp^2 carbon materials is due to the presence of oxygen-bearing carboxylic groups, formed at the edges and edge-like defects of graphene sheets. These oxygen-containing moieties are naturally present in graphene sheets due to their spontaneous air oxidation [2]. Pumera [21] employed XPS, cyclic voltammetry, and amperometry to provide evidence that the adsorption of NAD+ and passivation of electrodes occurs at the edges and edge-like defects of graphene. XPS and Car-Parrinello molecular dynamics revealed the following: When NAD+ was positioned close to a graphene sheet edge that contains a –COO– group, there was significant interaction that agreed well with the experimental results [2]. On the other hand, no such relevant interactions were found when NAD+ was located close to the basal plane of graphene or near hydrogen-only substituted edges of graphene sheets (Figure 3.8) [2,24].

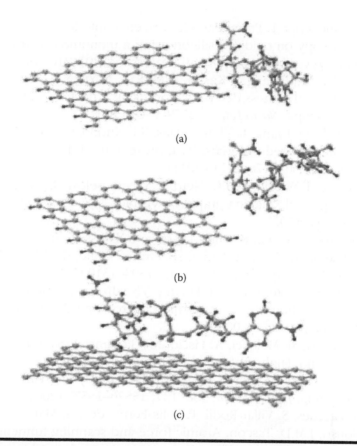

(a)

(b)

(c)

Figure 3.8 (See color insert.) Nicotine adenine adsorption onto the graphene sheets. Geometries for adsorption of NAD+ on a (a) model showing the H-atom and one –COO termination at the edge plane of graphene; (b) graphene edge plane totally terminated by hydrogen atoms; and (c) model of graphene sheet's basal plane using Car-Parrinello molecular dynamics. C, gray; N, blue; O, red; P, yellow; H, black. (Reproduced with permission from M. Pumera, R. Scipioni, H. Iwai, T. Ohno, Y. Miyahara, M. Boero, *Chem. Eur. J.* 15, 10851–10856, 2009.)

References

1. I. Jung, M. Pelton, R. Piner, D.A. Dikin, S. Stankovich, S. Watcharotone, et al., Simple approach for high-contrast optical imaging and characterization of graphene-based sheets, *Nano Lett.* 7, 3569–3575 (2007).

2. A. Lambacher, P. Fromherz, Fluorescence interference-contrast microscopy on oxidized silicon using a monomolecular dye layer, *Appl. Phys. A Mater. Sci. Process.* 63, 207–216 (1996).
3. Z.H. Ni, H.M. Wang, J. Kasim, H.M. Fan, T. Yu, Y.H. Wu, et al., Graphene thickness determination using reflection and contrast spectroscopy, *Nano Lett.* 7, 2758–2763 (2007).
4. V. Singh, D. Joung, L. Zhai, S. Das, S.I. Khondaker, S. Seal, Graphene based materials: Past, present and future, *Prog. Mater. Sci.* 56, 1178–1271 (2011).
5. P. Blake, E.W. Hill, A.H.C. Neto, K.S. Novoselov, D. Jiang, R. Yang, et al., Making graphene visible, *Appl. Phys. Lett.* 91, 063124 (2007).
6. J.S. Park, A. Reina, R. Saito, J. Kong, G. Dresselhaus, M.S. Dresselhaus, G' band Raman spectra of single, double and triple layer graphene, *Carbon* 47, 1303–1310 (2009).
7. J. Kim, L.J. Cote, F. Kim, J. Huang, Visualizing graphene based sheets by fluorescence quenching microscopy, *J. Am. Chem. Soc.* 132, 260–267 (2009).
8. E. Treossi, M. Melucci, A. Liscio, M. Gazzano, P. Samori, V. Palermo, High-contrast visualization of graphene oxide on dye-sensitized glass, quartz, and silicon by fluorescence quenching, *J. Am. Chem. Soc.* 131, 15576–15577 (2009).
9. J.I. Paredes, S. Villar-Rodil, P. Solis-Fernandez, A. Martinez-Alonso, J.M.D. Tascon, Atomic force and scanning tunneling microscopy imaging of graphene nanosheets derived from graphite oxide, *Langmuir* 25, 5957–5968 (2009).
10. C. Lee, X. Wei, J.W. Kysar, J. Hone, Measurement of the elastic properties and intrinsic strength of monolayer graphene, *Science* 321, 385–388 (2008).
11. C. Gómez-Navarro, J.C. Meyer, R.S. Sundaram, A. Chuvilin, S. Kurasch, M. Burghard, et al., Atomic structure of reduced graphene oxide, *Nano Lett.* 10, 1144–1148 (2010).
12. C.O. Girit, J.C. Meyer, R. Erni, M.D. Rossell, C. Kisielowski, L. Yang, et al., Graphene at the edge: Stability and dynamics, *Science* 323, 1705–1708 (2009).
13. J.C. Meyer, C. Kisielowski, R. Erni, M.D. Rossell, M.F. Crommie, A. Zettl, Direct imaging of lattice atoms and topological defects in graphene membranes, *Nano Lett.* 8, 3582–3586 (2008).
14. M.H. Gass, U. Bangert, A.L. Bleloch, P. Wang, R.R. Nair, A.K. Geim, Free-standing graphene at atomic resolution, *Nat. Nanotechnol.* 3, 676–681 (2008).

15. N.R. Wilson, P.A. Pandey, R. Beanland, R.J. Young, I.A. Kinloch, L. Gong, et al., Graphene oxide: structural analysis and application as a highly transparent support for electron microscopy, *ACS Nano* 3, 2547–2556 (2009).
16. J.C. Meyer, C.O. Girit, M.F. Crommie, A. Zettl, Imaging and dynamics of light atoms and molecules on graphene, *Nature* 454, 319–322 (2008).
17. R.R. Nair, P. Blake, J.R. Blake, R. Zan, S. Anissimova, U. Bangert, et al., Graphene as a transparent conductive support for studying biological molecules by transmission electron microscopy, *Appl. Phys. Lett.* 97, 3492845 (2010).
18. K. Erickson, R. Erni, Z. Lee, N. Alem, W. Gannett, A. Zettl, Determination of the local chemical structure of graphene oxide and reduced graphene oxide, *Adv. Mater.* 22, 4467–4472 (2010).
19. A.C. Ferrari, J.C. Meyer, V. Scardaci, C. Casiraghi, M. Lazzeri, F. Mauri, et al., Raman spectrum of graphene and graphene layers, *Phys. Rev. Lett.* 97, 187401 (2006).
20. I. Calizo, A.A. Balandin, W. Bao, F. Miao, C.N. Lau, Temperature dependence of the Raman spectra of graphene and graphene multilayers, *Nano Lett.* 7, 2645–2649 (2007).
21. M. Pumera, Electrochemistry of graphene: New horizons for sensing and energy storage, *Chem. Record* 9, 211–223 (2009).
22. T.J. Davis, M.E. Hyde, R.G. Compton, Nanotrench arrays reveal insight in to graphite electrochemistry, *Angew. Chem.* 117, 5251–5256 (2005).
23. A. Chou, T. Bocking, N.K. Singh, J.J. Gooding, Demonstration of the importance of oxygenated species at the ends of carbon nanotubes on their favorable electrochemical properties, *Chem. Commun.* 842–844 (2005).
24. M. Pumera, Electrochemical properties of double walled carbon nanotube electrodes, *Nanoscale. Res. Lett.* 2, 87–93 (2007).
25. X. Ji, C.E. Banks, A. Crossley, R.G. Compton, Oxygenated edge plane sites slow the electron transfer of the ferro/ferricyanide redox couple at graphite electrodes, *Chem. Phys. Chem.* 7, 1337–1344 (2006).
26. A.G. Crevillen, M. Pumera, M.C. Gonzalez, A. Escarpa, The preferential electrocatalytic behaviour of graphite and multiwalled carbon nanotubes on enediol groups and their analytical implications in real domains, *Analyst* 134, 657–662 (2009).

27. M. Pumera, R. Scipioni, H. Iwai, T. Ohno, Y. Miyahara, M. Boero, A mechanism of adsorption of β-nicotinamide adenine dinucleotide on graphene sheets: Experiment and theory, *Chem. Eur. J.* 15, 10851–10856 (2009).
28. C.M. Moore, S.D. Minteer, S.R. Martin, Microchip-based ethanol/oxygen biofuel cell, *Lab Chip* 5, 218–225 (2005).
29. N.G. Shang, P. Papakonstantinou, M. McMullan, M. Chu, A. Stamboulis, A. Potenza, S.S. Dhesi, H. Marchetto, Catalyst-free efficient growth, orientation and biosensing properties of multilayer graphene nano-flake films with sharp edge planes, *Adv. Funct. Mater.* 18, 3506–3514 (2008).
30. J. Lu, L.T. Drzal, R.M. Worden, I. Lee, Simple fabrication of a highly sensitive glucose biosensor using enzymes immobilized in exfoliated graphite nanoplatelets nafion membrane, *Chem. Mater.* 19, 6240–6246 (2007).
31. D. Kato, N. Sekioka, A. Ueda, R. Kurita, S. Hirono, K. Suzuki, O.J. Niwa, A nanocarbon film electrode as a platform for exploring DNA methylation, *J. Am. Chem. Soc.* 130, 3716–3717 (2008).

Chapter 4

Graphene-Based Materials in Gas Sensors

4.1 Graphene-Based Materials as Gas Sensors

Nanomaterials are widely preferred over other materials for gas sensing due to their unique and outstanding properties, such as extremely high surface-to-volume ratio [1]. The foretold property can potentially lead to novel sensors with exceptional performance while reducing the device size and minimizing energy consumption [1]. In addition, the electron transport through graphene is highly sensitive to the adsorbed molecules owing to the two-dimensional structure of graphene that makes every carbon atom a surface atom [1]. Graphene has been demonstrated as a promising gas-sensing material [1–3]; for instance, Sheehan et al. [3] reported that mechanically exfoliated graphene can potentially detect gaseous species down to the single molecular level. The gas-sensing mechanism of graphene is generally ascribed to the adsorption/desorption of gaseous molecules (which act as electron donors or acceptors) on the graphene surface, which leads to changes in the conductance of graphene [1,2].

Chen et al. [1] demonstrated the possibility of sensitive gas sensing with graphene materials. Chen et al. [1], designed their sensing device by dispersing the aqueous graphene oxide (GO) suspension onto gold interdigitated electrodes [4] with both finger-width and interfinger spacing (source-drain separation) of about 1 μm. These electrodes were fabricated using an e-beam lithography process on a silicon wafer with a top layer of thermally formed SiO_2 (200 nm). A few drops of the GO suspension were cast onto gold interdigitated electrodes, and a discrete network of GO sheets was left behind on the wafer after water evaporation. The working principle of the sensing device is that the drain-source channel becomes closed after GO is partially reduced by low-temperature thermal treatments; thus, the conductance of the device varies on exposure to various gases.

Chen et al. [1] carried out the thermal reduction of GO in a tube furnace (Lindberg Blue, TF55035A–1) by two modes: successive multistep heating and one-step heating. For successive multistep heating, three cycles on the order of 100, 200, and 300°C were performed on GO devices; the duration for each heating cycle was 1 h in an argon flow during the process. For one-step heating, GO devices were treated in the furnace at 200°C in an argon flow for 2 h. After heating, samples were quickly cooled to room temperature within 5 min (with the help of a blower). Both two-terminal direct current (DC) and three-terminal field effect transistor (FET) (Figure 4.1) measurements were performed on GO devices using a Keithley 2602 source meter. Electrical conductance of the GO device was measured by ramping the drain-source voltage Vds and simultaneously recording the drain-source current Ids to evaluate the influence of thermal treatment on the device characteristics. The bottom of the silicon wafer was used as the back-gate electrode during FET measurements.

Chen et al. [1] characterized the sensing performance of the as-fabricated GO devices under practical conditions (i.e., room temperature and atmospheric pressure) against low-concentration NO_2 and NH_3 diluted in dry air. An airtight test chamber with an electrical feed-through was used to house a GO device for gas-sensing characterizations [1,5]. The chamber

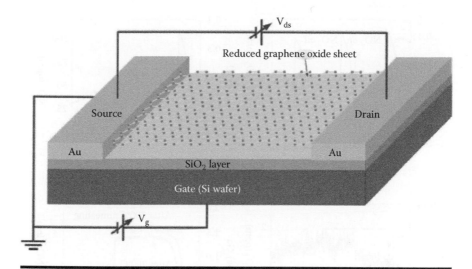

Figure 4.1 Schematic diagram of the reduced GO device. A reduced GO sheet bridges the source and drain electrodes, which closes the circuit. In FET measurements, the back of the silicon wafer is used as the gate electrode. (Reproduced with permission from G. Lu, L.E. Ocola, J. Chen, *Nanotechnol.* 20, 445502, 2009.)

volume (6.3×10^{-5} m^3) was minimized to reduce the capacitive effect. Variations in the electrical conductance of GO were monitored by simultaneously applying a low constant DC voltage (0.1 to 5 V) and recording the change in current passage through the device when the device was exposed periodically to clean air and air contaminated with NO_2^- or NH_3^- [1]. A sensing test cycle typically consisted of three consecutive steps, with exposures of the device to (1) clean airflow to record a base value of the sensor conductance, (2) target gas to register a sensing signal, and (3) clean airflow for sensor recovery [1].

Chen et al. [1] demonstrated that devices with as-deposited GO sheets showed no response to 100-ppm NO_2 or 1% NH_3, indicating insignificant change in the electrical transport property of the nonreduced GO. Further, they also found that 1-h, 100°C heating was usually inadequate to make GO devices responsive to gases. After being partially reduced by either successive multistep heating at 100 and 200°C (1 h each) or one-step heating at 200°C (2 h), the GO devices became

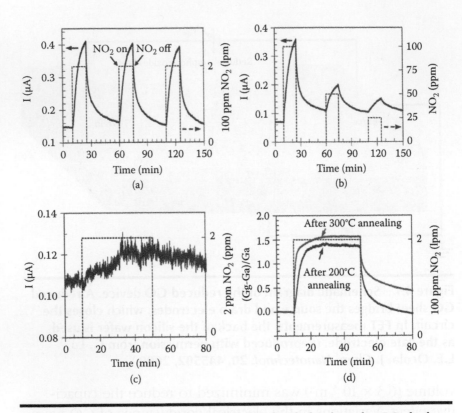

Figure 4.2   Room temperature NO₂ sensing behavior of a GO device thermally treated with successive multistep heating. After annealing in argon at 100 and 200°C for 1 h each, (a) the GO device showed a repeatable response to 100-ppm NO₂, (b) the sensing signal was highly dependent on the NO₂ concentration, and (c) the GO sensor could detect NO₂ at a concentration as low as 2 ppm. (d) Annealing at 300°C improved both sensor sensitivity and response time but lengthened the recovery time compared with 200°C annealing. (Reproduced with permission from G. Lu, L.E. Ocola, J. Chen, *Nanotechnol.* 20, 445502, 2009.)

highly responsive to NO₂ and NH₃, which was most likely due to the recovery of many graphitic carbon atoms as active sites for target gas adsorption [1]. Possibly, vacancies or small holes were created during thermal treatment, and these defects may also serve as adsorption sites for gaseous molecules [1,6].

Figure 4.2a shows a typical dynamic response (current vs. time) of a GO device for room temperature detection of

100-ppm NO_2 after being successively heated at 100 and 200°C in argon for 1 h each. The sensor was periodically exposed to clean dry airflow (2 liters per minute [LPM]) for 10 min to record a base value of the sensor conductance, 100-ppm NO_2 diluted in air (2 lpm) for 15 min to register a sensing signal, and clean airflow (2 lpm) again for 25 min to recover the device [1]. On the introduction of NO_2, the sensor current increased rapidly (i.e., the conductance of the sensor increased); when the NO_2 flow was turned off and the airflow restored, the device reestablished its conductance in about 30 min [1].

Three cycles were repeated (Figure 4.2a), and the signal was fairly reproducible. The sensing signal strength (proportional to the spike height with NO_2 on) was dependent on the NO_2 concentration as shown in Figure 4.2b, and it decreased with decreasing NO_2 concentrations from 100 to 50 and to 25 ppm [1]. Figure 4.2c shows the sensor response to 2-ppm NO_2; the conductance increased 12% with 40 min of NO_2 exposure. Assuming a linear relationship between conductance change and NO_2 concentration, this sensitivity is comparable to the sensor based on mechanically cleaved graphene, which showed a 4.3% increment in the conductance for 1-ppm NO_2 [1,2].

The sensing performance of the devices reported by Chen et al. [1] was encouraging for practical applications when considering the simplicity and low cost to fabricate these devices and the potential opportunities for optimization. The sensing performance was attributed to the effective adsorption of NO_2 on the surface of reduced, p-type GO [1]. NO_2 is a strong oxidizer and an electron-withdrawing group; therefore, electron transfer from reduced GO to adsorbed NO_2 leads to an enriched hole concentration and enhanced electrical conduction in the reduced GO sheet [1,7].

Figure 4.2d compares the sensitivities of the device for 100-ppm NO_2 detection after 200 and 300°C annealing. The sensor sensitivity was evaluated as the ratio of $(Gg - Ga)/Ga$, where Ga is the sensor conductance in clean air, and Gg is the sensor conductance in air containing 100-ppm NO_2. Chen et al. [1] found that the device showed a sensitivity of 1.56 to 100 ppm NO_2 after 300°C annealing, which is higher than that (1.41) after 200°C annealing. In addition, it had a faster response when exposed to NO_2, as evidenced by a steeper slope on the exposure to NO_2 [1]. This accelerated response was due to the creation of carbon atoms that were more graphitic during the 300°C annealing because molecular adsorption onto sp^2-bonded carbon (lower binding energy required) is faster than onto defects [1,3]. However, the sensor recovery after 300°C annealing became slower because the device did not return to its initial conductance after 30 min exposure to dry air, whereas in the 200°C annealing case, the full recovery was achieved under the same exposure condition [1]. Low-temperature heating and ultraviolet (UV) illumination could be used to accelerate the sensor recovery [1].

The GO devices responded to NH_3 as well, after either successive thermal treatments or one-step heating. Figure 4.2 represents the 1% NH_3 sensing data obtained from the same reduced GO device (after 300°C heating), whose NO_2 sensing behavior is shown in Figure 4.2. On NH_3 exposure, the current passing through the device decreased (Figure 4.2a) due to the adsorption of NH_3 (an electron donator), which lowered the hole concentration in GO, thereby reducing the GO conductance [1]. Unlike the relatively fast recovery (about 30 min) after NO_2 exposure, the conductance of the GO device could not return to its initial value in airflow (2 lpm), even after 50 h (shown in Figure 4.2b).

Chen et al. [1] provided evidence that most of the GO sensors (with one exception to be discussed further) recovered slowly after NH_3 sensing. However, further investigation is essential to understand the intrinsic mechanisms associated with the slow recovery of reduced GO from NH_3 exposure and to find effective measures to accelerate the recovery process before reduced GO can be used practically for repeatable NH_3 detection [1]. Among all the GO devices fabricated

(11 total), the device developed by Chen et al. [1] is the only device that recovered exceptionally fast after NH_3 sensing.

Figure 4.3a shows three NH_3 sensing cycles using this device after it was reduced through one-step heating in argon at 200°C for 2 h [1]. This NH_3 sensing performance is superior to that shown in Figure 4.3a; however, an unusual current increase was noticed at the beginning of NH_3 exposure in each cycle, as indicated by the arrows in Figure 4.3a, which is against the GO sensing mechanism discussed previously and was not observed with other devices [1]. Figure 4.3b shows a magnified view of the area marked by the leftmost arrow in Figure 4.3a, indicating that the current increased with a sharp slope at the beginning of NH_3 exposure [1]. This abnormal increase may be attributed to the unsteadiness of flow field in the chamber during gas switching or some other accidental noise [1].

However, even stranger behavior was noticed in this device when employed to sense 1% NH_3 after 2 months [1]. As shown in Figure 4.3c, the device gave a sensing signal (curve I) with a completely opposite trend (i.e., its conductance increased on NH_3 exposure) after 2 months, compared with that obtained freshly after reduction (curve II, obtained by normalizing the curve in Figure 4.3a). The current I is normalized to the initial value I_0 in airflow for the convenience of comparison. This observation by Chen and coworkers [1] suggested that competing sensing mechanisms may exist besides gas adsorption/desorption on reduced p-type GO for this device.

The device was then re-treated in argon at 200°C for 2 h; although later it was restored to "normal" sensing behavior (conductance decreased on NH_3 exposure), as shown by curve III in Figure 4.3c, its sensitivity was degraded [1]. Figure 4.3d represents the *Ids-Vds* curve of the device measured 2 months after the first thermal treatment [1]. The curve is asymmetric and nonlinear, which is contrary to the one acquired shortly after the thermal treatment (inset of Figure 4.3d) and suggests a nonohmic contact between the GO and gold electrodes. This contact issue may have played a role in the abnormal NH_3 sensing

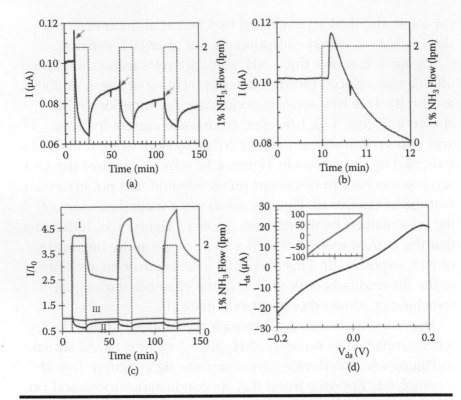

Figure 4.3 (a) Detection of 1% NH_3 by a GO device treated with one-step heating (in argon at 200°C for 2 h). This device recovered faster compared with the one presented in Figure 4.5. (b) On exposure to NH_3, the conductance of the device first oddly increased for a few seconds (this increase cannot be explained by a NH_3-adsorption-induced change in GO conductance) before it decreased (as expected). (c) The sensing response for 1% NH_3 after 2 months (curve I) is opposite (i.e., the conductance increased on NH_3 exposure) to that observed after freshly heated (curve II, obtained by normalizing the curve in Figure 4.3a). The sensor response returned to normal after another heating treatment (in argon at 200°C for 2 h), but the sensitivity degraded, as indicated by curve III. (d) The I_{ds}-V_{ds} curve of the device became asymmetric and nonlinear, implying non-ohmic contact between the GO and gold electrodes; the inset shows the I_{ds}-V_{ds} obtained fresh after the first heating. (Reproduced with permission from G. Lu, L.E. Ocola, J. Chen, *Nanotechnol.* 20, 445502, 2009.)

behavior. For instance, Peng et al. proposed that the Schottky barrier modulation at the contact of the carbon nanotube (CNT) with metal significantly contributed to NH_3 sensing when using CNT sensors at room temperature [1,8]. Therefore, the influence of graphene-metal contact on the sensor performance is worthy of future study. Chen et al. [1] could not exclude the connection between the abnormal NH_3 sensing behavior and the unusually fast recovery. Understanding this odd NH_3 sensing behavior may lead to measures that can be used to engineer sensing device properties, such as recovery rate and gas selectivity [1].

4.1.1 Improving Graphene's Gas Sensing by the Insertion of Dopants or Defects

To completely exploit the possibilities of graphene sensors, it is important to understand the interaction between the graphene surface and the adsorbed molecules [9]. Theoretical studies have been conducted to investigate the adsorption of small molecules on graphene. Most of the earlier work focused on perfect graphene and predicted relatively low adsorption energies in comparison with the essential requirement of gas-sensing applications [9–13]. In practice, the graphene sheets prepared by the available fabrication methods are likely to have many defects [9]. Besides, graphene could be deliberately or accidentally doped by noncarbon elements [9]. To date, the number of studies on the effects of dopants and defects on the sensing properties of graphene is surprisingly small.

In a recent work, Zhang et al. [9] reported a first-principle simulation of the interactions between several small molecules and various graphene sheets. The model systems were carefully chosen to cover several basic issues [9]. The selected gas molecules included CO, NO, NO_2, and NH_3, which are all of great practical interest for industrial, environmental, and medical applications [9]. Meanwhile, NO_2 and NH_3 represent typical electron acceptors and donors, which may undergo charge transfer with graphene [9]. The graphenes are doped by boron and nitrogen atoms, representing the most widely used p- and n-type dopants. For the defective graphene, only one defect containing a single missing atom in each supercell is considered for minimizing the complexity. Structurally perfect graphene is also studied for comparison [9]. Zhang et al. [9] performed this work to understand the fundamental insights on the influence of adsorbed molecules on the electronic properties of graphene. Further, another reason for this study is to understand how these effects could be utilized to design gas-sensing devices that are more sensitive.

Zhang et al. [9] performed the density functional theory (DFT) calculations using CASTEP [14], an ultrasoft pseudopotential, plane-wave basis, and periodic boundary conditions. The local density approximation (LDA) with CA-PZ (Ceperley-Alder, Perdew-Zungar) functional and a 240-eV cutoff energy for the plane-wave basis set were used in all relaxation processes [9]. Each simulated system under investigation consisted of a 12.30 × 12.30 × 10 Å graphene supercell (50 carbon atoms) with a single molecule adsorbed in the central region (Figure 4.4). The distance between adjacent graphene layers was kept at 10 Å [9]. The k-point was set to 3 × 3 × 1 for the Brillouin zone integration. The structural configurations of the isolated graphenes were optimized through fully relaxing the atomic structures [9]. With the same supercell and k-point samplings, the configurations of the molecule-graphene systems were optimized through fully relaxing the atomic structures until the remaining forces were smaller than 0.01 eVÅ^{-1} [9]. The adsorption energy of the small molecules on graphene was calculated as

$$E_{ad} = E_{(molecule+graphene)} - E_{(graphene)} - E_{(molecule)} \tag{4.1}$$

where $E_{(molecule+graphene)}$, $E_{(graphene)}$, and $E_{(molecule)}$ are the total energies of the relaxed molecule on the graphene system, graphene, and the molecule, respectively. For the density-of-state (DOS) calculation, the k-point was set to 9 × 9 × 1 to achieve high accuracy.

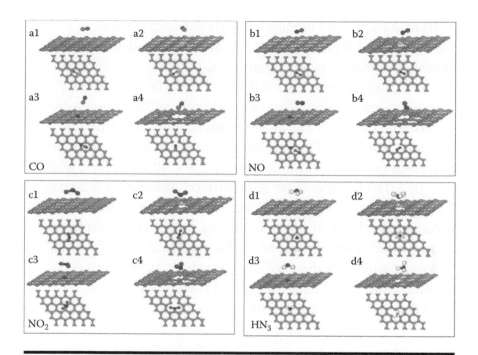

Figure 4.4 **Schematic view of the favorable adsorption configurations of the CO, NO, NOB2B, and NHB3B molecules on the (a1, b1, c1, d1) P-, (a2, b2, c2, d2) B-, (a3, b3, c3, d3) N-, and (a4, b4, c4, d4) D-graphenes. Carbon and boron atoms are shown as gray; hydrogen atoms are shown in white, and the nitrogen and oxygen atoms are shown in black. (Reproduced with permission from Y.-H. Zhang, Y.-B. Chen, K.-G. Zhou, C.-H. Liu, J. Zeng, H.-L. Zhang, Y. Peng,** *Nanotechnology* **20, 185504, 2009.)**

Zhang et al. [9] performed the electron transport calculations using the Atomistix ToolKit (ATK) 2.0.4 package [15], which implements a DFT-based, real-space, nonequilibrium Green's function (NEGF) formalism [16–21]. The mesh cutoff was chosen as 200 Ryd to achieve a reasonable balance between the calculation efficiency and accuracy. The current was calculated by the Landauer-Buttiker equation [9]. For simplicity, the pristine, boron-doped, nitrogen-doped, and defective graphenes were denoted as P-graphene, B-graphene, N-graphene, and D-graphene, respectively, in their study [9].

To find the most favorable adsorption configurations, Zhang et al. [9] initially placed the molecule under investigation at different positions above a graphene sheet with different orientations. After full relaxation, the optimized configurations obtained from the different initial states were compared to identify the most energetically stable one [9]. The most stable configurations of the CO, NO, NO_2, and NH_3 molecules on the pristine (P-), boron (B-), nitrogen (N-), and defective (D-) graphene are summarized in Figure 4.4. More detailed information from the simulation of the different molecule-graphene systems, including values of adsorption energy, equilibrium graphene-molecule distance (defined as the center-to-center distance of the nearest atoms between graphene and small molecules), and the charge transfer (Mulliken charge), are listed in Table 4.1.

4.1.1.1 CO on Graphene

Zhang et al. [9] considered several initial configurations to study the adsorption of CO on the P-graphene. A CO molecule was initially placed above a carbon atom or the center of a six-member ring (6MR), with the CO molecule oriented perpendicular (with the carbon or oxygen atom pointing toward the graphene sheet) to the graphene [9]. Several other configurations with the CO molecule placed parallel to the graphene plane were also tested by Zhang et al. [9]. After full relaxation, a configuration with the adsorbed CO axis aligned parallel to the graphene plane along the axis of two opposite carbon atoms of the 6MR was found to be the most stable one for the P-graphene [9]. The adsorption energy of this system was −0.12 eV, and the molecule-sheet distance was estimated as 3.02 Å (Figure 4.4a1). The low adsorption energy and long distance indicate a weak interaction.

The charge transfer between CO and P-graphene was obtained from Mulliken population analysis (Table 4.1). For CO on P-graphene, the calculated charges on the carbon and

Table 4.1   Adsorption Energy E_{ad}, Equilibrium Graphene-Molecule Distance d (Defined as the Shortest Atom-to-Atom Distance), and Mulliken Charge Q of Small Molecules Adsorbed on Different Graphene Sheets

System	E_{ad}	d (A)	Q (e)[a]
CO on P-graphene	−0.12	3.02	−0.01
NO on P-graphene	−0.30	2.43	0.04
NO$_2$ on P-graphene	−0.48	2.73	−0.19
NH$_3$ on P-graphene	−0.11	2.85	0.02
CO on B-graphene	−0.14	2.97	−0.02
NO on B-graphene	−1.07	1.99	0.15
NO$_2$ on B-graphene	−1.37	1.67	−0.34
NH$_3$ on B-graphene	−0.50	1.66	0.40
CO on N-graphene	−0.14	3.15	0
NO on N-graphene	−0.40	2.32	0.01
NO$_2$ on N-graphene	−0.98	2.87	−0.55
NH$_3$ on N-graphene	−0.12	2.86	0.04
CO on D-graphene	−2.33	1.33	0.26
NO on D-graphene	−3.04	1.34	−0.29
NO$_2$ on D-graphene	−3.04	1.42	−0.38
NH$_3$ on D-graphene	−0.24	2.61	0.02

[a] Q is defined as the total Mulliken charge on the molecules, and a negative number means charge transfer from graphene to molecule.

oxygen atoms of the CO were $0.42|e|$ and $−0.43|e|$, respectively, while there was no charge on the carbon atoms of the P-graphene. Nevertheless, a very small charge $(0.01|e|)$ was transferred from the P-graphene to CO [9]. When adsorbed on B-graphene, CO adopted a tilted orientation with respect to the plane of the B-containing 6MR, with the carbon atom close to the boron atom [9]. The adsorption energy and charge transfer of the CO on the B-graphene system were found to be -0.14 eV and $-0.02|e|$, respectively [9]. The CO on N-graphene system showed a similar adsorption energy of -0.14 eV to the CO on B-graphene system with no charge transfers. The adsorption energies values suggest that CO cannot distinguish p- and n-type dopants on graphene, which is different from the previous report on CO-nanotube interaction [9,19]. For the D-graphene, the configuration with CO was tilted with respect to the graphene plane, and the carbon atom pointing toward the vacancy was found to be the most favorable one [9].

The calculation indicated that the graphene carbon atom close to the vacancy defect provided a stronger binding site for the CO molecule than those further away from the vacancy [9]. The minimum atom-to-atom distance between the CO and the D-graphene was found to be 1.33 Å [9]. This distance was, in fact, close to the bond length of a C–C double bond and was much shorter than that of the other three types of graphenes, which were 3.02 Å (P-graphene), 2.97 Å (B-graphene), and 3.15 Å (N-graphene) (Figures 4.4a1–4.4a3), respectively [9]. The adsorption energy of CO on the D-graphene can reach -2.33eV, which was more than one magnitude higher than that of the pristine and doped graphene [9]. The electronic total charge density plot for the CO on P-graphene was compared with that of the CO on D-graphene, as shown in Figure 4.5. No electron orbital overlap between the CO molecule and the P-graphene was observed in the CO on the P-graphene system in Figure 4.5a. In contrast, Figure 4.5b shows that the electronic charge plot of CO and the D-graphene were strongly overlapped, leading to more orbital mixing and a larger charge transfer. The electronic total charge density plots for the CO on B-graphene and CO on N-graphene (not shown) indicated similar features as that of the CO on P-graphene, in which no electron density overlap was observed [9].

The total charge density analysis illustrated that only weak physisorption took place between the CO and P-, N-, and

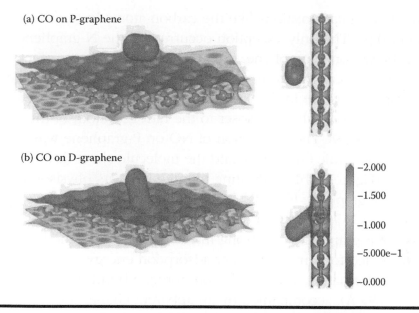

(a) CO on P-graphene

(b) CO on D-graphene

−2.000
−1.500
−1.000
−5.000e−1
−0.000

Figure 4.5 Electronic total charge densities for the adsorption adducts of (a) CO on P-graphene and (b) CO on D-graphene. (Reproduced with permission from Y.-H. Zhang, Y.-B. Chen, K.-G. Zhou, C.-H. Liu, J. Zeng, H.-L. Zhang, Y. Peng, *Nanotechnology* 20, 185504, 2009.)

B-graphenes, while the vacancy on D-graphene provided strong chemisorption binding sites for the CO [9]. The strong orbital overlap between CO and D-graphene resulted in a significant change to the electronic properties of the graphene, and Zhang et al. [9] proposed that the D-graphene is expected to be more suitable for sensing CO than the P-, B-, and N-graphenes.

4.1.1.2 NO on Graphene

Similarly, Zhang et al. [9] placed the NO molecule on various sites of the four graphenes with different orientations to find the optimal adsorption configurations. The favorable configurations of NO on the different graphenes were similar to the CO on graphene systems (i.e., the nitrogen atom adopted similar positions like the carbon atoms in the CO system) [9]. The only exception occurred on the N-graphene. For the CO on N-graphene, the CO molecule took an orientation with its carbon atom pointing at the nitrogen atom of the N-graphene, while in the NO on N-graphene, the oxygen atom in the NO molecule was closer to the nitrogen atom of the N-graphene [9]. The adsorption of NO on P-graphene was the least exothermic (−0.30 eV), and the molecule-sheet distance is 2.43 Å (Figure 4.4b1), indicating that the NO was physisorbed on the P-graphene [9]. This result is similar to the recent reports of NO adsorption on carbon nanotubes [9,20,21]. In the case of B-graphene, the strong interaction between the boron and NO led to a much stronger adsorption energy (−1.07 eV) and the formation of a tight boron-nitrogen bond (bond distance 1.99 Å), accompanied by an apparent charge transfer of $0.15|e|$ from NO to graphene sheet [9]. For the N-graphene, the adsorption energy was −0.40 eV, and the closest distance was 2.32 Å [9]. The D-graphene showed the highest affinity to NO, which gave a −3.04-eV adsorption energy, and the NO-graphene distance was only 1.34 Å [Figure 4.4(b4)], revealing the occurrence of strong chemisorption [9].

4.1.1.3 NO_2 on Graphene

Various possible configurations of the triangular NO_2 molecule adsorbed onto the graphene sheets were investigated by Zhang et al. [9]. This investigation was performed to understand the interaction between NO_2 and different types of graphene. Three major possible adsorption configurations were studied, similar to the previous studies on the adsorption of NO_2 on carbon nanotubes [22], including that the NO_2 molecule bonded to the sheet surface with a nitrogen end (referred to as the nitro configuration), bonded through one oxygen end (referred to as the nitrite configuration), and bonded with both oxygen ends (referred to as the cycloaddition configuration) [9]. The cycloaddition configuration on P-graphene gave rise to an adsorption energy of −0.48 eV, which was higher than the adsorption energy of the nitro configuration (−0.39 eV) or the nitrite configuration (−0.45 eV). The results indicated that the cycloaddition configuration favored the interaction between the electron-rich oxygen atoms and the carbon atoms on the graphene [9]. Meanwhile, a large charge transfer $(0.19|e|)$ from the graphene to NO_2 was observed, confirming that the NO_2 acted as an electron acceptor. The calculated adsorption energy (−0.48 eV) was in good agreement with the experimentally determined physisorption energy (−0.40 eV) [23] and the adsorption energy (−0.50 eV) for NO_2 on carbon nanotubes in theory [9,21]. On the B-graphene, the nitro configuration gave a stronger interaction than the other configurations. The interaction between the boron and nitrogen atom led to a high adsorption energy (−1.37 eV) and the formation of a tight boron-nitrogen bond (bond distance 1.67 Å, accompanied by an apparent charge transfer of $0.34|e|$ from the B-graphene to NO_2 [9]. The nitro configuration was also the most favorable one for both the N- and D-graphenes, giving adsorption energies of −0.98 and −3.04 eV, respectively [9].

4.1.1.4 NH_3 on Graphene

The NH_3 molecule showed different adsorption configurations on different graphenes, with a more complicated adsorption mechanism than the other molecules studied. On the P-graphene, the configuration with the three hydrogen atoms of NH_3 pointing toward the graphene plane was the favorable one (Figure 4.4d1), which gave an adsorption energy of −0.11 eV [9]. This result was consistent with previous reports about NH_3 adsorbed on carbon nanotubes (−0.14 eV) and NH_3 adsorbed on graphene (0–0.17 eV) [10,13], suggesting a weak interaction between NH_3 and the P-graphene [9]. On the B-graphene, NH_3 was attached to the boron atom with the nitrogen atom pointing at the sheet, which gave an adsorption energy of −0.50 eV and a boron-nitrogen distance of

1.66 Å (Figure 4.4d2). The adsorption configuration of NH_3 on the N-graphene was similar to that on the P-graphene, which had the hydrogen atoms pointing toward the sheet [9]. However, in the NH_3 on N-graphene system, the N atom was positioned above the nitrogen on the N-graphene, while in the NH_3 on the P-graphene system, the nitrogen was positioned above the 6MR center [9].

The calculated adsorption energy of NH_3 on the N-graphene was −0.12 eV, indicating the weak physisorption nature [9]. The adsorption of NH_3 on the D-graphene was slightly stronger, with an adsorption energy of −0.24 eV and little charge transfer [9]. The adsorption energy of NH_3 on the B-graphene (−0.50 eV) was much higher than that on the other three graphenes, which can be attributed to the strong interaction between the electron-deficient boron atom and the electron-donating nitrogen atom of NH_3 [9]. It was also found that the B-graphene underwent an obvious distortion on NH_3 adsorption (Figure 4.4d2), indicating that the boron site was transformed from sp^2 hybridization to sp^3 hybridization [9,10]. The boron-nitrogen distance (1.66 Å) was close to the boron-nitrogen bond length in BH_3NH_3 (1.6576 Å) [9,24], confirming the formation of a covalent bond between the NH_3 and the B-graphene. This strong interaction was also evident in the electronic total charge density of the NH_3 on the B-graphene system, which showed large electron density (not shown here). The interaction between adsorbed molecules and graphenes was expected to alter the electronic structure of the graphenes, which could be reflected by the change in electrical conductance of the graphene [9]. Strong interaction could bring about significant conductivity change, which is beneficial for sensing applications [9]. The calculation results suggest that the P-graphene had weak interactions with all four gas molecules [9]. Introducing dopants and defects into the graphene significantly increased the molecule-graphene interaction. Based on the analysis, Zhang et al. [9] predicted that the B- and D-graphenes are more suitable for gas-sensing applications since they have stronger interactions with the four small molecules than the P- and N-graphenes. More specifically, Zhang et al. [9] indicated that the D-graphene showed the highest sensitivity toward CO, NO, and NO_2, while B-graphene is the best choice for sensing NH_3.

4.1.2 Density of States of the Molecule–Graphene System

To verify the effects of the adsorption of small molecules on the graphene's electronic properties, the total electronic DOSs of the molecule-graphene adsorption systems were calculated, and a few DOSs for some representative systems are shown in Figure 4.6 [9]. The calculations performed on the adsorption energies have suggested that the CO on P- and B-graphenes and NH_3 on N-graphene systems have weak interactions between the molecules and the graphenes, and these weak interactions were also evident in their DOS structures (Figures 4.6a–4.6c), which showed little change after the adsorption [9]. For example, the DOSs of CO on P-graphene and CO on B-graphene were identical to those of the P- and B-graphenes, respectively [9]. The contribution of the CO electronic levels to the total DOS for both systems was localized between −10.0 and −2.6 eV in the valence bands and at 2.5 eV in the conduction bands, which are far away from the Fermi level [9].

Similarly, the contribution of the NH_3 electronic level in the NH_3 on N-graphene was localized at −2.3 eV (valence bands) and 2.5 eV (conduction bands), which are also far away from the Fermi level [9]. In contrast, Figures 4.6d–4.6f show that the DOSs of CO on D-graphene, NO on B-graphene, and NO_2 on N-graphene were drastically changed from those of the corresponding graphenes due to strong molecule-graphene interactions. On comparison, the DOS of D-graphene showed a larger peak than P-graphene, appearing just above the Fermi level [9]. This peak indicates the strong metallic nature of

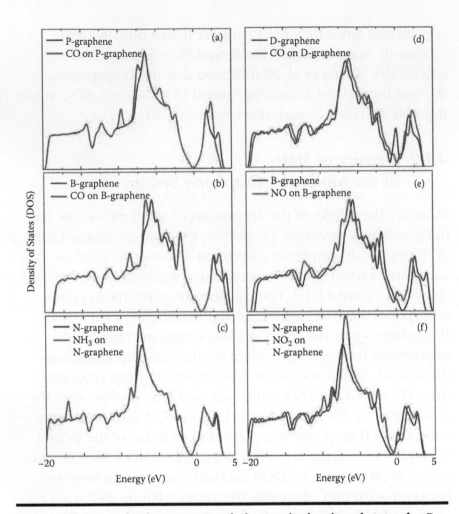

Figure 4.6 (See color insert.)   Total electronic density of states for P-, B-, N-, and D-graphene (black curves) and molecule-graphene systems (red curves) calculated for the corresponding configurations shown in Figure 4.4a1, 4.4a2, 4.4d3, 4.4a4, 4.4b2, and 4.4c3. The Fermi level was set to zero. (Reproduced with permission from Y.-H. Zhang, Y.-B. Chen, K.-G. Zhou, C.-H. Liu, J. Zeng, H.-L. Zhang, Y. Peng, *Nanotechnology* **20, 185504, 2009.)**

the system and a significant increase in conductivity compared to the P-graphene [9].

After the chemisorption of CO molecules, the system became more semiconductor-like, with a drop in the DOS near the Fermi level [9]. Consistent with the adsorption energy values, the DOS analysis also indicated that the interaction between CO and the D-graphene was stronger than that with the pristine one [9]. Such an enhancement in interaction can be directly associated with the rearrangement of the defective sheet structure in the presence of the CO [9]. Note that the adsorption of CO onto the D-graphene caused the major band features to move toward higher energy; in other words, the Fermi level was shifted toward lower energy [9]. The adsorption of NO onto B-graphene caused a notable increase of the DOS in the region just above the Fermi level, which was also expected to increase the conductance [9]. Meanwhile, the Fermi level shifted slightly toward higher energy after the adsorption. It is understood that the B-dopant introduced electronic holes into the graphene, which generated a p-type semiconductor [9]. When B-graphene interacted with an electron-rich NO molecule, a large charge transfer to the B-graphene occurred, which dramatically enhanced the conductivity of the NO on the B-graphene system [9]. For the NO_2 on N-graphene, the strong interaction caused a dramatic increase of the DOSs on both sides near the Fermi level. The change in the DOSs, especially the area near the Fermi level, was expected to bring about obvious changes in the corresponding electronic properties [9]. Therefore, Zhang et al. [9] concluded from Figure 4.6 that D-, B-, and N-graphene are suitable for sensing applications for CO, NO, and NO_2, respectively.

4.1.3 The I–V Curves of Molecules on Graphene

Zhang et al. [9] quantitatively evaluated the gas-sensing properties of the graphenes by simulating the electron transport properties of different graphenes using NEGF methods. The simplest type of chemical-sensing transducer is a resistance sensor, in which the resistance of the sensing materials on the adsorption of chemicals is detected [9]. Graphene-based resistance sensors are simulated using a model consisting of

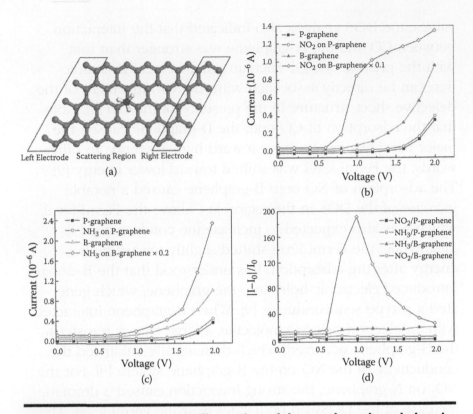

Figure 4.7 (a) A schematic illustration of the graphene-based chemical sensor for detecting small gas molecules. (b) A comparison of the *I-V* curves of the devices based on P-graphene, NO₂ on P-graphene, B-graphene, and NO₂ on B-graphene. (c) The *I-V* curves of the devices based on P-graphene, NH₃ on P-graphene, B-graphene, and NH₃ on B-graphene. (d) The normalized *I-V* curves of the P-graphene and B-graphene. It should be noted that the *I-V* curves in Figure 4.7b and 4.7c are offset 0.02 × 10⁻⁶ A from each other, while the curves in Figure 4.7d are offset from each other for clarity. (Reproduced with permission from Y.-H. Zhang, Y.-B. Chen, K.-G. Zhou, C.-H. Liu, J. Zeng, H.-L. Zhang, Y. Peng, *Nanotechnology* 20, 185504, 2009.)

a graphene sheet contacted by two graphene electrodes, as shown in Figure 4.7a.

Zhang et al. [9] calculated a series of current-versus-voltage (*I-V*) curves for such graphene junctions with and without the adsorption of different molecules [9]. The simulated *I-V* curves for the P-graphene and B-graphene before and after NO₂ and

NH_3 adsorption are illustrated in Figures 4.7b and 4.7c. The *I-V* curve of the P-graphene exhibited a nonlinear behavior. Further, it was observed that the B-graphene was about three times more conductive than the P-graphene due to the increased hole-type charge carriers, which confirmed Zhang et al.'s [9] previous finding in the DOS analysis (Figure 4.6F). NO_2 adsorption on graphene resulted in a slight current increase in the conductivity [9]. In contrast, a dramatic increase of current was observed for the NO_2 on B-graphene, indicating a much higher sensitivity. After normalizing against the intrinsic conductivity of the corresponding graphenes (Figure 4.7d), the sensitivity of B-graphene to NO_2 was found to be dependent on the bias voltage, and a high sensitivity bias window between 0.8 and 1.2 V was visible [9]. At the optimum bias voltage of 1.0 V, the B-graphene showed a sensitivity nearly two orders of magnitude higher than that of the P-graphene (Figure 4.7d). The B-graphene was less sensitive to NH_3 than to NO_2 but still showed one order of magnitude higher sensitivity than the P-graphene when the bias voltage was higher than 1.0 V [9].

4.1.4 Concerns for Practical Application

Although the results of the calculation discussed suggest that the doped and defective graphenes exhibited improved sensing properties compared to the pristine graphene, it is worth noting then the strong binding between the modified graphenes and certain molecules may also bring about some serious drawbacks [9]. For example, strong binding implies that desorption of gas molecules from the doped and defective graphenes could be difficult, and the devices may suffer from longer recovery times. Novoselov et al. [2] demonstrated that a graphene sensor could be regenerated to its initial state within 100–200 s by annealing at 150°C in vacuum or short UV irradiation. However, a much longer recovery time is expected if the adsorption energy is significantly increased. According to the conventional transition state theory, the recovery time τ can be expressed as

$$\tau = v_0^{-1} e^{(-EB/K_BT)}$$

where T is the temperature, K_B is the Boltzmann's constant, and v_0 is the attempt frequency. Increasing the adsorption energy E_{ad} will prolong the recovery time in an exponential manner. Zhang et al. [9] estimated that, in a strong binding case, like NO_2 on D-graphene, the recovery time could be on the order of 1010 h at 600 K, which is obviously not acceptable for any practical application.

The desorption of the small molecules from the graphene surface may be assisted by UV irradiation [9,25] or electric field [9,26], which has been exploited to clean up adsorbents on carbon nanotubes or metals. But, the effectiveness of these cleaning methods has not been fully investigated on graphene devices [9]. Therefore, before any innovative cleaning method is uncovered, the graphene-based sensing devices are more likely to find applications as highly sensitive irreversible sensors rather than ideal reversible sensors [9]. A graphene-based sensor is still in its early stages, and much work is needed before it may compete with the many currently available sensors [9]. A fundamental understanding of the binding phenomena of small molecules on graphene is essential to explore this new field [9].

4.2 Graphene as a Membrane for Gas Separation

Jiang et al. [27] demonstrated that graphene can be employed as membranes for gas separation. Membrane separation of gases is affordable due to its low energy cost [28]. For example, the purification and production of H_2 by the most common route of methane re-forming requires the separation of H_2 from other gases [29]. Various membranes have been developed for hydrogen separation, including silica, zeolite, carbon-based, and polymer membranes. However, the thickness of these membranes varies from tens of nanometers to several micrometers. Because the permeance of a membrane is inversely proportional to the thickness of the membrane [30], these membranes may be limited in their overall efficiency or productivity. As a result, the one-atom-thick graphene nanosheet was chosen as the ultimate membrane. Understanding molecular and atomic transport through such a truly two-dimensional membrane is not only fundamentally interesting but also useful in a number of applications, such as proton exchange membranes in fuel cells, separating gases to increase the sensitivity of chemical sensors and the separation of CO_2 from the exhaust gases of industrial/power plants.

The perfect graphene sheet, however, is impermeable to gases as small as helium [31]. This is because the electron density of its aromatic rings is high enough to repel atoms and molecules trying to pass through these rings. Therefore, to achieve gas permeability, it is necessary to drill holes in the graphene sheet. Recently, the electron beam of a transmission electron microscope has been employed to punch closely spaced nanopores within suspended graphene sheets [32].

Alternatively, molecular building blocks have been assembled to create porous two-dimensional sheets [33]. Improvements in these techniques may be fruitful for creating ordered subnanometer-size pores within graphene sheets that may be used in the future as two-dimensional molecular sieve membranes for gas separation.

Sint et al. [34] employed classical molecular dynamics (MD) simulations to study the diffusion rates of solvated ions such as Na^+, K^+, Cl^-, and Br^- passing through graphene nanopores driven by an external electric field. These studies highlighted the possibility of employing graphene as ion separation membranes as they were able to simulate pores that exhibited selectivity toward either cations or anions. According to Sint et al. [34], the charges on the graphene sheet and their responses to the ions/molecules passing through were not straightforward to manage in classical MD simulations but would be automatically accounted for in the framework of first-principles DFT. More important, gas separation, is important to the chemical industry and has the ability to save energy. The separation of gases through graphene has not been explored previously. To the best of knowledge, Jiang et al. [27] were the first to demonstrate the membrane separation capability of a porous graphene sheet for molecular gases (H_2 vs. CH_4) by designing subnanometer-size pores and modeling their selective diffusion of gas molecules using first-principles methods.

Further, Jiang et al. [27] employed DFT calculations with plane-wave bases and periodic boundary conditions to explore the potential energy surface (PES) and dynamics of hydrogen and methane molecules passing through subnanometer pores created in a graphene sheet. Jiang et al. [27] performed the initial static calculations using both the Perdew, Burke, and Erzenhoff (PBE14) functional form of the generalized gradient approximation (GGA) and the Rutgers-Chalmers van der Waals density functional (vdW-DF) for exchange and correlation [35,36]. Dispersion interactions should be important for the interaction of neutral, nonpolar molecules such as H_2 and CH_4 with the aromatic rings of the graphene sheet. Therefore, Jiang et al. [27] employed the vdW-DF to evaluate the strength of these interactions. The vdW-DF has been extensively tested on numerous systems [37], including the physisorption of small molecules to graphene sheets [38,39] and the adsorption of H_2 within metal-organic framework materials [40]. Jiang et al. [27] also employed vdW-DF15 as implemented in a modified version of the Abinit [41] plane wave, norm-conserving pseudopotentials, and a kinetic energy cutoff of 680 eV for plane waves. Further, in the large unit cell required to model the porous graphene, only the Γ-point was used for sampling the Brillouin zone.

Jiang et al. [27] performed the first-principles MD (FPMD) simulations using the Vienna ab initio simulation package [42,43] and the all-electron projector-augmented wave method [44,45] within the frozen core approximation to describe the electron-core interaction, thus requiring a lower kinetic energy cutoff (300 eV in this case) [45,46]. For the FPMD simulations, H_2 or CH_4 molecules were placed within a hexagonal cell with the porous graphene in the *ab*-plane and a *c*-dimension of 12 Å. Constant-temperature simulations with a time step of 1 fs at 600 K were performed. Figure 4.8a illustrates the creation of a pore that displays high H_2/CH_4 selectivity. Jiang et al. [27] first removed two neighboring rings from a graphene sheet with a 6 × 6 hexagonal lattice that had a periodicity of 1.47 nm (see Figure 4.8b) based on the PBE-optimized lattice parameter for graphene (2.45 Å). Next, four unsaturated carbon atoms were passivated with hydrogen atoms; the remaining four were substituted with nitrogen atoms (Figure 4.8a).

Figure 4.9 depicts the electron density isosurface of the graphene pore at a rather low value of 0.02 e/Å^3. Here, the pore can be seen to be approximately rectangular with dimensions of 3.0 × 3.8 Å. Certainly, an experimental realization of this porous graphene may be challenging; however, Jiang et al. [27] proposed two promising approaches based on their experimental efforts. In the first approach, Jiang et al. [27] employed an electron beam to punch holes in the graphene sheet [32], and the dangling bonds were passivated by nitrogen doping with ammonia [46]. In the second approach, Jiang et al. [27] synthesized a building block similar to the one enclosed in the dotted lines in Figure 4.8b and then assembled the building blocks into a larger sheet of porous graphene. This approach was successfully applied to construct a two-dimensional framework from building blocks created via the trimerization of terephthalonitrile [33]. The resulting porous material consisted of 1.5-nm channels, featuring many pyridinic nitrogen atoms around the pore rim [33]. Although refinements in these techniques will be necessary to create well-ordered,

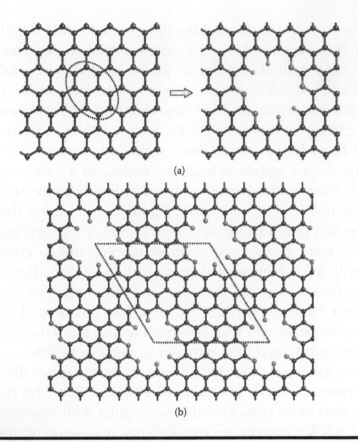

Figure 4.8 (See color insert.) **(a) Creation of a nitrogen-functionalized pore within a graphene sheet: The carbon atoms in the dotted circle are removed, and four dangling bonds are saturated by hydrogen; the other four dangling bonds together with their carbon atoms are replaced by nitrogen atoms. (b) The hexagonally ordered porous graphene. The dotted lines indicate the unit cell of the porous graphene. Carbon, black; nitrogen, green; hydrogen, cyan. (Reproduced with permission from D. Jiang, V.R. Cooper, S. Dai, *Nano Lett.* 9, 4019–4024, 2009.)**

appropriate-size pores, these successes offer potential pathways for synthesizing the porous graphene proposed in Figure 4.8.

Jiang et al. [27] examined several adsorption configurations for H_2 placed at the center of the pore. Energetics from PBE calculations showed that both the out-of-plane orientation and the in-plane orientation with the H_2 bond axis pointing toward the passivating hydrogen atoms were slightly repulsive, while

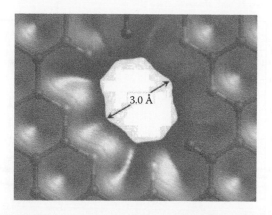

Figure 4.9 Pore electron density isosurface of the nitrogen-functionalized porous graphene (isovalue of 0.02 e/Å³). (Reproduced with permission from D. Jiang, V.R. Cooper, S. Dai, *Nano Lett.* 9, 4019–4024, 2009.)

the in-plane orientation with the H_2 bond axis pointing toward nitrogen atoms was slightly attractive (Jiang et al. [27] denoted this orientation as $XYZH_2$; see Figure 4.10 inset). vdW-DF yielded the same energetic ordering for the three orientations. Jiang et al. [27] attributed the favorable interaction of H_2 at the $XYZH_2$ configuration to the attraction between H_2 and the nitrogen atoms.

Following this, Jiang et al. [27] then explored the PES for displacing the H_2 molecule from the $XYZH_2$ position perpendicularly out of the graphene plane. Figure 4.10 shows the PES for both PBE and vdW-DF. Both methods indicated that the PES was relatively flat (with barriers of 0.025 and 0.04 eV for PBE and vdW-DF, respectively) for H_2 moving in and out of the pore. Including the vdW interaction shifts, the PBE PES decreased by about 0.05 eV, although not uniformly. vdW-DF showed that H_2 interacted most strongly at 1.6 Å above the pore. It should be noted that the true adsorption height may be 0.25–0.35 Å smaller as vdW-DF has a tendency to overestimate separation distances [47] at room temperature, which corresponds to an energy of 0.025 eV. As a result, Jiang et al. [27] expected that the H_2 molecule would be able to overcome the barrier to passing through this pore. The observed result was

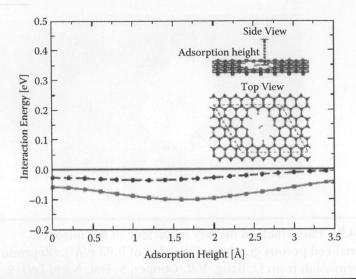

Figure 4.10 (See color insert.) **Interaction energy between H₂ and the nitrogen-functionalized porous graphene as a function of adsorption height. Insets show the definition of adsorption height and orientation of H₂ in the pore. Red squares, solid lines, vdW-DF; black circles, dashed lines, PBE. (Reproduced with permission from D. Jiang, V.R. Cooper, S. Dai, *Nano Lett.* 9, 4019–4024, 2009.)**

consistent with the fact that the kinetic radius of H₂ (2.89 Å) is smaller than the width of the pore (3.0 Å).

To investigate the selectivity of the pore for the diffusion of H₂ relative to CH₄, Jiang et al. [27] explored the PES for CH₄ passing through the pore. Similar to the H₂ calculations, Jiang et al. [27], examined several configurations of CH₄ adsorbed in the pore center and found that in the most stable orientation, the four hydrogen atoms of CH₄ pointed toward the four corners of the rectangular pore (see inset of Figure 4.11).

In this configuration, PBE showed a repulsive interaction of 0.41 eV. The inclusion of vdW interactions gave a slightly less-repulsive energy of 0.33 eV. The PES for displacing the CH₄ molecule out of the pore showed a parabolic decrease in the repulsive interaction that, after an inflection point at about 1 Å, reached a shallow attractive well at larger separation distances for both PBE and vdW-DF. In the latter case, the well was

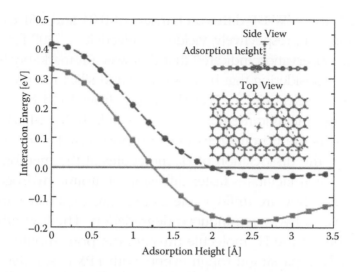

Figure 4.11 (See color insert.) Interaction energy between CH_4 and the nitrogen-functionalized porous graphene as a function of adsorption height. Insets show the definition of adsorption height and orientation of CH_4 in the pore. Red squares, solid lines, vdW-DF; black circles, dashed lines, PBE. (Reproduced with permission from D. Jiang, V.R. Cooper, S. Dai, *Nano Lett.* 9, 4019–4024, 2009.)

deeper, with a depth of −0.18 eV around 2.5 Å; for the former, the well was shallow, with a depth of −0.03 eV around 2.75 Å. Incorporation of vdW interactions shifted the PES curve down by an average 0.1 eV, and one can consider the PBE curve to sufficiently approximate the shape of the vdW-DF PES. This barrier was also consistent with the larger kinetic radius of CH_4 (at 3.8 Å) relative to the pore width. On the basis of molecular diffusion barriers, the pore selectivity of H_2/CH_4 can be easily estimated. In contrast with traditional membranes for H_2 separation, such as silica and polymers, for which selectivity is a product of diffusivity and solubility ratios, the concepts of solubility and free volume are not applicable for the one-atom-thin graphene membrane. Here, the diffusivity ratio alone determines the selectivity of the gas molecule passing through the pore. Assuming that the diffusion rates follow the Arrhenius rate equation and that the prefactor were of the same magnitude for both gases [27], the vdW-DF diffusion barriers of 0.04 and 0.51 eV for H_2 and CH_4, respectively, yielded a selectivity of 108 for H_2/CH_4. The observation indicated that this was a remarkably high selectivity considering that both polymer and silica membranes usually have selectivity for H_2/CH_4 ranging from 10 to 10^3.

Although metallic membranes have similarly high H_2/CH_4 selectivity as the porous graphenes proposed by Jiang et al. [27] due to the dissociation of H_2 molecules at the membrane surface, which facilitates faster diffusion of atomic hydrogen species [29], they are unfavorable as they are expensive and susceptible to hydrogen-induced degradation. The low diffusion barrier of the H_2 molecule through the pore should allow us to observe passing-through events with FPMD simulations.

Given that PBE and vdW-DF give similar PES for H_2 interacting with the pore and that the vdW-DF method has yet to be incorporated within an MD simulation code, Jiang et al. [27] employed the PBE method for the MD simulations. Moreover, Jiang et al. [27] performed an NVT simulation at 600 K. Jiang et al. [27] chose the high temperature to accelerate the dynamics to observe passing-through events within a reasonable simulation time. During a 36-ps run, Jiang et al. [27] observed an H_2 passing-through rate of 0.1 molecules/ps. This relatively fast rate was simply a reflection of the relatively low potential energy barrier for H_2 passing through the pore, as indicated by the PES. Figure 4.12 shows snapshots of a passing-through event. At 244 fs, the H_2 molecule entered the pore in an $XYZH_2$ position, as predicted by the geometry optimization discussed previously, and then stayed there for about 180 fs. At 424 fs, the H_2 molecule began to diffuse out of the pore.

Similar MD simulations for the CH_4 molecule were performed by Jiang et al. [27], but no passing-through events were observed for the same time frame. Hence, the FPMD simulations further illustrated the high permselectivity of the porous graphene for H_2/CH_4 separation. Using our MD simulation for H_2 passing through the pore, Jiang et al. [27] gave a crude estimate of the H_2 flux through the porous graphene membrane.

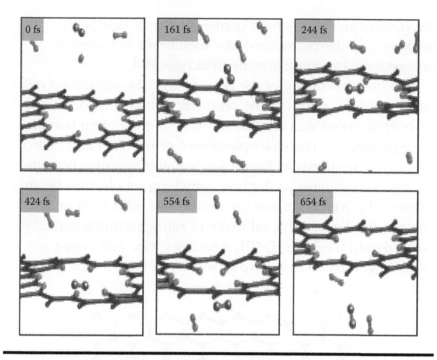

Figure 4.12 Snapshots of H$_2$ diffusing through the nitrogen-functionalized pore from first principles molecular dynamics simulations at 600 K. The passing-through H$_2$ molecule is highlighted in dark gray. (Reproduced with permission from D. Jiang, V.R. Cooper, S. Dai, *Nano Lett.* **9, 4019–4024, 2009.)**

Averaging the number of passing-through events over the simulation time (36 ps) and taking into account the area of the membrane (187 Å^2), a flux of 10 mol H$_2$ cm^{-2} s^{-1} results. Assuming a pressure drop of 1 bar across the membrane, we arrive at an H$_2$ permeance of 1 mol m^{-2} s^{-1} Pa^{-1}. If the under-estimate (by ~ 0.015 eV) of the diffusion barrier by PBE were considered, the permeance would be lowered by a factor of only 1.3 at 600 K. For comparison, a 30-nm-thick silica membrane has an H$_2$ permeance of 5 × 10^{-7} mol m^{-2} s^{-1} Pa^{-1} at 673 K [27], and a similarly thick silica-alumina membrane has an H$_2$ permeance of 2–3 × 10^{-7} mol m^{-2} s^{-1} Pa^{-1} at 873 K [27]. Polymeric membranes tend to have even lower H$_2$ permeance [27] than oxide-based membranes. Jiang et al. [27], attributed the porous graphene's high permeance to its one-atom thickness, as the permeance of a membrane is inversely proportional to the membrane thickness [30].

The high selectivity of the pore for H_2/CH_4 separation relies on the precise control of the pore size, which was determined by how the pore was created and how the dangling bonds were passivated. The exact placement of nitrogen and hydrogen atoms as shown in Figure 4.8 probably requires bottom-up organic synthesis as the hole-punching by electron beam followed by NH_3 treatment are unlikely to offer such precise control. To alleviate this difficulty of nitrogen functionalization but maintain the high H_2/CH_4 selectivity, one can create an all-hydrogen passivated pore as shown in Figure 4.13. In this

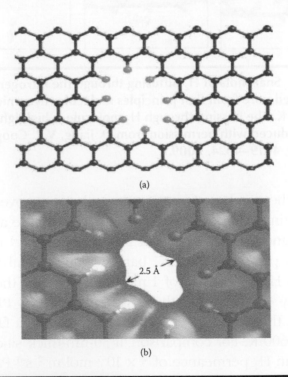

(a)

2.5 Å

(b)

Figure 4.13 An all-hydrogen passivated pore in graphene (a) and the pore electron-density isosurface (b). Isovalue is at 0.02 e/Å³. (Reproduced with permission from D. Jiang, V.R. Cooper, S. Dai, *Nano Lett.* 9, 4019–4024, 2009.)

case, the eight dangling σ-bonds are saturated with hydrogen rather than substituting four carbon atoms with nitrogen. The width of the pore was narrowed to 2.5 Å due to the additional hydrogen atoms, while the length of the pore remained at about 3.8 Å (Figure 4.13b). This smaller pore size did have an effect on the diffusion barriers: Jiang et al. [27] found that the vdW-DF barrier for H_2 and CH_4 increased to 0.22 and 1.60 eV, respectively (Figures 4.14 and 4.15, respectively) [27]. Since the selectivity was dictated by the ratio of the two diffusion barriers, the dramatically increased CH_4 diffusion barrier coupled with the small change in the H_2 diffusion barriers translated into a drastic increase in the H_2/CH_4 selectivity in the all-hydrogen pore relative to that of the nitrogen-functionalized pore. Again, using the Arrhenius equation to estimate the selectivity, this new pore will have a selectivity of H_2/CH_4 at about 10^{23} at room temperature.

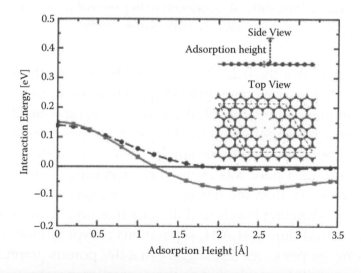

Figure 4.14 (See color insert.) Interaction energy between H_2 and the all-hydrogen passivated porous graphene as a function of adsorption height. Insets show the definition of adsorption height and orientation of H_2 in the pore. Red squares, solid lines, vdW-DF; black circles, dashed lines, PBE. (Reproduced with permission from D. Jiang, V.R. Cooper, S. Dai, *Nano Lett.* 9, 4019–4024, 2009.)

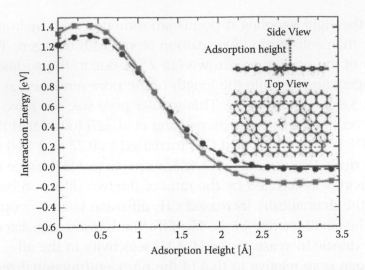

Figure 4.15 (See color insert.) Interaction energy between CH₄ and the all-hydrogen passivated porous graphene as a function of adsorption height. Insets show the definition of adsorption height and orientation of CH₄ in the pore. Red squares, solid lines, vdW-DF; black circles, dashed lines, PBE. (Reproduced with permission from D. Jiang, V.R. Cooper, S. Dai, *Nano Lett.* 9, 4019–4024, 2009.)

Also, note that the 0.22-eV barrier for H₂ passing through the pore can be overcome quite frequently at room temperature: Assuming a prefactor of about 10^{13} s^{-1}, Jiang et al. [27] estimated a passing-through frequency of 10^9 per second at room temperature. Due to its simpler chemical construction, this all-hydrogen pore was even more promising for separating H₂ from CH₄ than the nitrogen-functionalized pore. The absence of dangling bonds and the fact that no chemical reactions occurred during H₂/CH₄ separations at typical moderate operating temperatures [29] suggest that the porous graphene membrane will be stable during the separation process.

A porous graphene membrane, like other membranes for gas separation, will be susceptible to short-circuiting by large-size pore defects. To assess this effect, Jiang et al. [27] initially examined two larger pore sizes to determine at what size the pore would lose its H₂/CH₄ selectivity. Using DFT-PBE calculations, Jiang et al. [27] found that H_2 could pass through the two larger pores without a barrier (not shown), while the diffusion barrier for CH_4 decreased to a negligible 0.02 eV for the medium-size pore (width at 3.8 Å) and to zero for the largest pore (5.0 Å). For pores greater than 3.8 Å (which is roughly the kinetic radius of CH_4), both H_2 and CH_4 diffused through without a barrier.

Jiang et al. [27] then analyzed the graphene membrane's selectivity as a function of the concentration of larger-pore defects, assuming a simple two-pore-size model in which the smaller pore was highly selective for H_2/CH_4, while the larger pore allowed both gases to pass without a barrier. Moreover, Jiang et al. [27] found that for a given temperature, there was a critical concentration of larger-pore defects for a desired selectivity (not shown). Furthermore, increasing temperature can be used to suppress the effect of defects (not shown). This analysis was useful for understanding the capacity of porous graphene for gas separation, given the reality of imperfect manufacturing techniques. For separation of small gas molecules, the key parameter demonstrated here was the pore width with respect to the kinetic radii of the target molecules. Although the two types of pores examined in Figures 4.9 and 4.13 were similar in structure, they had different widths (2.5 vs. 3.0 Å), which made a big difference in separating H_2 (kinetic radius 2.9 Å) from CH_4 (kinetic radius 3.8 Å). On the basis of these calculations, Jiang et al. [27] concluded that, for a given separation, the design principle is to construct a pore with size close to or slightly smaller than that of the smaller molecule of the gas mixture. To fine-tune the pore size, one can consider different passivating groups for the pore rim (such as hydrogen vs. nitrogen). It is certainly challenging to make subnanometer pores in graphene, as the smallest pores achieved experimentally so far are still in the range of several nanometers [32, 33]. However, Jiang et al. [27] hope that their work will stimulate experimentalists to take up this challenge.

References

1. G. Lu, L.E. Ocola, J. Chen, Reduced graphene oxide for room-temperature gas sensors, *Nanotechnol.* 20, 445502 (2009).
2. F. Schedin, A.K. Geim, S.V. Morozov, E.W. Hill, P. Blake, M.I. Katsnelson, K.S. Novoselov, Detection of individual gas molecules adsorbed on graphene, *Nat. Mater.* 6, 652–655 (2007).
3. J.T. Robinson, F.K. Perkins, E.S. Snow, Z.Q. Wei, P.E. Sheehan, Reduced graphene oxide molecular sensors, *Nano Lett.* 8, 3137–3140 (2008).
4. G.H. Lu, K.L. Huebner, L.E. Ocola, M. Gajdardziska-Josifovska, J.H. Chen, Gas sensors based on tin-oxide nanoparticles synthesized from a mini-arc plasma source, *J. Nanomater.* 2006, 60828 (2006).
5. G.H. Lu, L.E. Ocola, J.H. Chen, Room-temperature gas sensing based on electron transfer between discrete tin oxide nanocrystals and multiwalled carbon nanotubes, *Adv. Mater.* 21, 2487–2491 (2009).
6. I. Jung, D. Dikin, S. Park, W. Cai, S.L. Mielke, R.S. Ruoff, Effect of water vapor on electrical properties of individual reduced graphene oxide sheets, *J. Phys. Chem.* C 112, 20264–20268 (2008).
7. O. Leenaerts, B. Partoens, F.M. Peeters, Adsorption of H_2O, NH_3, CO, NO_2, and NO on graphene: A first-principles study, *Phys. Rev. B* 77, 125416 (2008).
8. N. Peng, Q. Zhang, C.L. Chow, O.K. Tan, N. Marzari, Sensing mechanisms for carbon nanotube based NH_3 gas detection, *Nano Lett.* 9, 1626–1630 (2009).
9. Y.-H. Zhang, Y.-B. Chen, K.-G. Zhou, C.-H. Liu, J. Zeng, H.-L. Zhang, Y. Peng, Improving gas sensing properties of graphene by introducing dopants and defects: A first-principles study, *Nanotechnology* 20, 185504 (2009).
10. L. Bai, Z. Zhou, Computational study of B- or N-doped single-walled carbon nanotubes as NH_3 and NO_2 sensors, *Carbon* 45, 2105–2110 (2007).
11. S.B. Fagan, A.G.S. Filho, J.O.G. Lima, J.M. Filho, O.P. Ferreira, I.O. Mazali, O.L. Alves, M. Dresselhaus, 1,2-Dichlorobenzene interacting with carbon nanotubes, *Nano Lett.* 4, 1285–1288 (2004).
12. Y. Zhang, D. Zhang, C. Liu, Novel chemical sensor for cyanides: Boron-doped carbon nanotubes, *J. Phys. Chem. B* 110, 4671–4674 (2006).

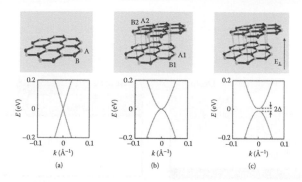

(a)

(b)

(c)

Figure 1.2 Band gap in graphene. Schematic diagrams of the lattice structure of (a) monolayer and (b) bilayer graphene. The green and red lattice sites indicate the A (A1/A2) and B (B1/B2) atoms of monolayer (bilayer) graphene, respectively. The diagrams represent the calculated energy dispersion relations in the low-energy regime and show that monolayer and bilayer graphene are zero-gap semiconductors. (c) When an electric field E is applied perpendicular to the bilayer, a band gap is opened in bilayer graphene, whose size (2Δ) is tunable by the electric field. (Reproduced with permission from J.B. Oostinga, H.B. Heersche, X. Liu, A.F. Morpurgo, L.M.K. Vandersypen, *Nat. Mater.* 7, 151–157, 2008.)

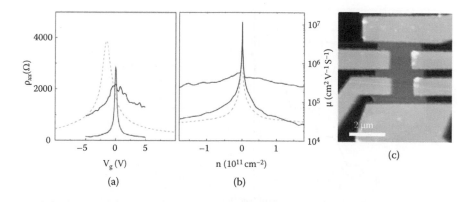

(a)

(b)

Figure 1.3 (a) Four-probe resistivity ρ_{xx} as a function of gate voltage V_g before (blue) and after (red) current annealing; data from a traditional high-mobility device on the substrate (gray dotted line) is shown for comparison. The gate voltage is limited to the ±5-V range to avoid mechanical collapse. (b) Mobility, $\mu = 1/en\,\rho_{xx}$ as a function of carrier density n for the same devices. (c) Atomic force microscopy (AFM) image of the suspended graphene before the measurements. (Reproduced with permission from K.I. Bolotin, K.J. Sikes, Z. Jiang, M. Klima, G. Fudenberg, J. Hone, et al., *Solid State Commun.* 146, 351–355, 2008.)

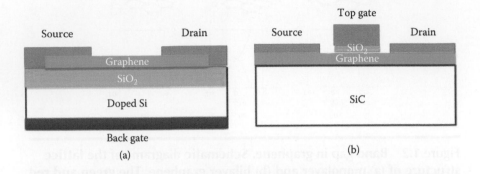

Figure 1.4 (a) Schematic diagram of back-gated and (b) top-gated graphene field effect devices. The perpendicular electric field is controllable by applied back-gate voltage V_g and the top-gate voltage V_{top}. (Reproduced with permission from V. Singh, D. Joung, L. Zhai, S. Das, S. I. Khondaker, S. Seal, *Prog. Mater. Sci.* 56, 1178–1271, 2011.)

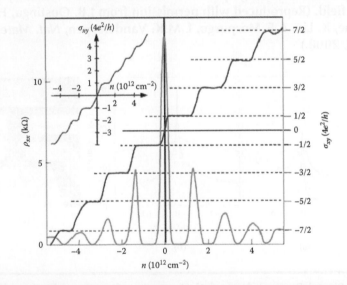

Figure 1.5 QHE for massless Dirac Fermions. Hall conductivity σ_{xy} and longitudinal resistivity ρ_{xx} of graphene as a function of their concentration at B = 14T and T = 4K. $\sigma_{xy} \equiv (4e^2/h)\nu$ is calculated from the measured dependencies of $\rho_{xy}(v_g)$ and $\rho_{xx}(v_g)$ as $\rho_{xy} = \rho_{xy}/(\rho^2_{xy}+\rho^2_{xx})$. The behavior of $1/\rho_{xy}$ is similar but exhibits a discontinuity at $v_g \approx 0$, which is avoided by plotting σ_{xy}. Inset: σ_{xy} in two layer graphene where the quantization sequence is normal and occurs at integer v. The latter shows that the half-integer QHE is exclusive to ideal-graphene. (Reproduced with permission from *Nature* 438, 197–200, 2005.)

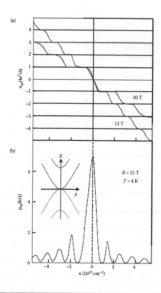

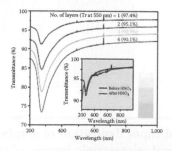

Figure 1.6 Quantum Hall effect in bilayer graphene. (a) Hall conductivity σ_{xy} and (b) longitudinal ρ_{xx} are plotted as functions of n at a fixed B and temperature $T = 4$ K. σ_{xy} allows the underlying sequences of QHE plateaus to be seen more clearly. *rxy* crosses zero without any sign of the zero-level plateau as expected in a conventional 2D system. The inset shows the calculated energy spectrum for bilayer graphene, which is parabolic at low e. (Reproduced with permission from K.S. Novoselov, E. McCann, S.V. Morozov, V.I. Fal'ko, M.I. Katsnelson, U. Zeitler, et al., *Nat. Phys.* 2, 177–180, 2006.)

Figure 1.7 Representative of transmittance of different graphene layers. UV-vis spectra roll-to-roll, layer-by-layer transferred graphene films on quartz substrates. The inset shows the UV spectra of graphene films with and without HNO$_3$ doping. (Reproduced with permission from S. Bae, H. Kim, Y. Lee, X. Xu, J.-S. Park, Y. Zheng, et al., *Nat. Nanotechnol.* 5, 574–578, 2010.)

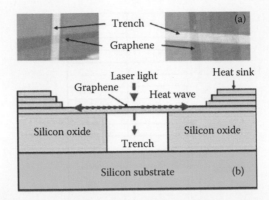

Figure 1.8 (a) High-resolution scanning electron microscopic image of the suspended graphene flakes. (b) Schematic of the experimental setup for measuring the thermal conductivity of graphene. (Reproduced with permission from S. Ghosh, I. Calizo, D. Teweldebrhan, E.P. Pokatilov, D.L. Nika, A.A. Balandin, et al., *Appl. Phys. Lett.* 92, 151911, 2008.)

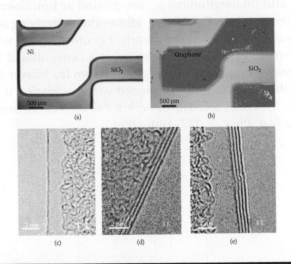

Figure 2.10 Chemical vapor deposition of graphene on transition metal substrates. Optical microscopic images of (a) the nickel catalyst and (b) the resulting graphene film. TEM images show the nucleation of (c) one, (d) three, or (e) four layers during the growth process. (Reproduced with permission from A. Reina, X.T. Jia, J. Ho, D. Nezich, H.B. Son, V. Bulovic, M.S. Dresselhaus, J. Kong, *Nano Lett.* 9, 30–35, 2009. Copyright 2009 American Chemical Society.)

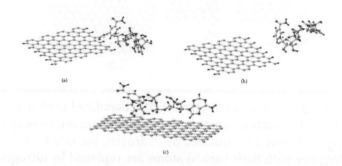

Figure 3.8 Nicotine adenine adsorption onto the graphene sheets. Geometries for adsorption of NAD⁺ on a (a) model showing the H-atom and one –COO termination at the edge plane of graphene; (b) graphene edge plane totally terminated by hydrogen atoms; and (c) model of graphene sheet's basal plane using Car-Parrinello molecular dynamics. C, gray; N, blue; O, red; P, yellow; H, black. (Reproduced with permission from M. Pumera, R. Scipioni, H. Iwai, T. Ohno, Y. Miyahara, M. Boero, *Chem. Eur. J.* 15, 10851–10856, 2009.)

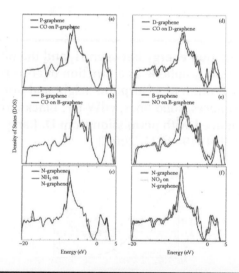

Figure 4.6 Total electronic density of states for P-, B-, N-, and D-graphene (black curves) and molecule-graphene systems (red curves) calculated for the corresponding configurations shown in Figure 4.4a1, 4.4a2, 4.4d3, 4.4a4, 4.4b2, and 4.4c3. The Fermi level was set to zero. (Reproduced with permission from Y.-H. Zhang, Y.-B. Chen, K.-G. Zhou, C.-H. Liu, J. Zeng, H.-L. Zhang, Y. Peng, *Nanotechnology* 20, 185504, 2009.)

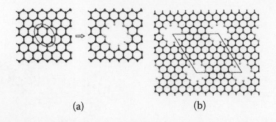

(a)   (b)

Figure 4.8   (a) Creation of a nitrogen-functionalized pore within a graphene sheet: The carbon atoms in the dotted circle are removed, and four dangling bonds are saturated by hydrogen; the other four dangling bonds together with their carbon atoms are replaced by nitrogen atoms. (b) The hexagonally ordered porous graphene. The dotted lines indicate the unit cell of the porous graphene. Carbon, black; nitrogen, green; hydrogen, cyan. (Reproduced with permission from D. Jiang, V.R. Cooper, S. Dai, *Nano Lett.* 9, 4019–4024, 2009.)

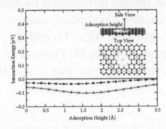

Figure 4.10   Interaction energy between H_2 and the nitrogen-functionalized porous graphene as a function of adsorption height. Insets show the definition of adsorption height and orientation of H_2 in the pore. Red squares, solid lines, vdW-DF; black circles, dashed lines, PBE. (Reproduced with permission from D. Jiang, V.R. Cooper, S. Dai, *Nano Lett.* 9, 4019–4024, 2009.)

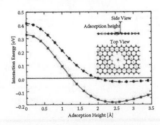

Figure 4.11   Interaction energy between CH_4 and the nitrogen-functionalized porous graphene as a function of adsorption height. Insets show the definition of adsorption height and orientation of CH_4 in the pore. Red squares, solid lines, vdW-DF; black circles, dashed lines, PBE. (Reproduced with permission from D. Jiang, V.R. Cooper, S. Dai, *Nano Lett.* 9, 4019–4024, 2009.)

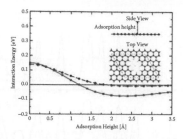

Figure 4.14   Interaction energy between H_2 and the all-hydrogen passivated porous graphene as a function of adsorption height. Insets show the definition of adsorption height and orientation of H_2 in the pore. Red squares, solid lines, vdW-DF; black circles, dashed lines, PBE. (Reproduced with permission from D. Jiang, V.R. Cooper, S. Dai, *Nano Lett.* 9, 4019–4024, 2009.)

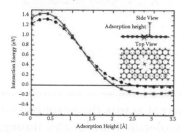

Figure 4.15   Interaction energy between CH_4 and the all-hydrogen passivated porous graphene as a function of adsorption height. Insets show the definition of adsorption height and orientation of CH_4 in the pore. Red squares, solid lines, vdW-DF; black circles, dashed lines, PBE. (Reproduced with permission from D. Jiang, V.R. Cooper, S. Dai, *Nano Lett.* 9, 4019–4024, 2009.)

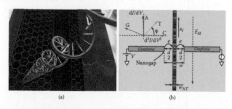

Figure 5.3   (a) The individual bases of the ssDNA molecule (backbone in green, bases in alternating colors) sequentially occupy a gap in graphene (hexagonal lattice) while translocating through it. Their conductance is read, revealing the sequence of the molecule. The contacting electrodes to the graphene nanogap (gold, yellow) are on the far left and right sides of this image. (b) Schematic representation of the transverse conductance technique. (Reproduced with permission from H.W. Ch. Postma, *Nano. Lett.* 10, 420–425, 2010.)

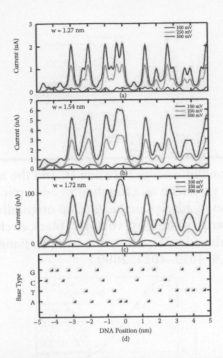

Figure 5.4 (a)–(c) Current across the graphene nanogap while a single-stranded DNA molecule translocates through it for three different gap widths *w* and bias voltage levels as indicated. (d) For this simulation, a random sequence of CGG CGA GTA GCA TAA GCG AGT CAT GTT GT was used. (Reproduced with permission from H.W. Ch. Postma, *Nano. Lett.* 10, 420–425, 2010.)

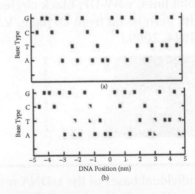

Figure 5.5 (a) The angle is used to determine the base type as described in the text (blue triangles, base type; red triangles, deduced base type from ψ). For *w* = 1.1–1.6 nm, the base type was deduced accurately. (b) When *w* = 1.7 nm, the overlapping current peaks caused misidentification. (Reproduced with permission from H.W. Ch. Postma, *Nano. Lett.* 10, 420–425, 2010.)

13. W. Charles, J. Bauschlicher, A. Ricca, Binding of NH_3 to graphite and to a (9,0) carbon nanotube, *Phys. Rev.* B 70, 115409 (2004).
14. V. Milman, B. Winkler, J.A. White, C.J. Pickard, M.C. Payne, E.V. Akhmatskaya, R.H. Nobes, Electronic structure, properties and phase stability of inorganic crystals: The pseudopotential plane-wave approach, *Int. J. Quantum Chem.* 77, 895–910 (2000).
15. J. Taylor, H. Guo, J. Wang, Ab initio modeling of quantum transport properties of molecular electronic devices, *Phys. Rev. B* 63, 245407 (2001).
16. M. Brandbyge, J.-L. Mozos, P. Ordejon, J. Taylor, K. Stokbro, Density functional method for non-equilibrium electron transport, *Phys. Rev. B* 65, 165401 (2002).
17. J.M. Soler, E. Artacho, J.D. Gale, A. García, J. Junquera, P. Ordejon, D. Sanchez-Portal, The SIESTA method for ab initio order-N materials simulation, *J. Phys.: Condens. Matter* 14, 2745 (2002).
18. M. Brandbyge, N. Kobayashi, M. Tsukada, Conduction channels at finite bias in single-atom gold contacts, *Phys. Rev. B* 60 17064 (1999).
19. S. Peng, K. Cho, Ab initio study of doped carbon nanotube sensor, *Nano Lett.* 3, 513–517 (2003).
20. R. Wang, D. Zhang, W. Sun, Z. Han, C. Liu, A novel aluminum doped carbon nanotube sensor for carbon monoxide, *J. Mol. Struct.: Theochem.* 806, 93–97 (2007).
21. S. Peng, K. Cho, P. Qi, H. Dai, Ab initio study of CNT NO_2 sensor, *Chem. Phys. Lett.* 387, 271–276 (2004).
22. W.L. Yim, X.G. Gong, Z.F. Liu, Chemisorption of NO_2 on carbon nanotubes, *J. Phys. Chem. B* 107, 9363 (2003).
23. P. Sjovall, S.K. So, B. Kasemo, R. Franchy, W. Ho, NO_2 adsorption on graphite at 90K, *Chem. Phys. Lett.* 172, 125–130 (1990).
24. D.A. Dixon, M. Gutowski, Thermodynamic properties of molecular borane amines and the [BH4-][NH4+] salt for chemical hydrogen storage systems from ab-initio electronic structure theory, *J. Phys. Chem. A* 109, 5129–5135 (2005).
25. R. Chen, N. Franklin, J. Kong, J. Cao, T. Tombler, Y. Zhang, H. Dai, Molecular photodesorption from single-walled carbon nanotubes, *Appl. Phys. Lett.* 79, 2258 (2001).
26. M.P. Hyman, J.W. Medlin, Theoretical study of the adsorption and dissociation of oxygen on Pt(111) in the presence of homogeneous electric fields, *J. Phys. Chem. B* 109, 6304–6310 (2005).

27. D. Jiang, V.R. Cooper, S. Dai, Porous graphene as ultimate membrane for gas separation, *Nano Lett.* 9, 4019–4024 (2009).

28. M. Freemantle, Advanced organic and inorganic materials being developed for separations offer cost benefits for environmental and energy-related processes, *Chem. Eng. News* 83, 49–57 (2005).

29. N.W. Ockwig, T.M. Nenoff, Membranes for hydrogen separation, *Chem. Rev.* 107, 4078–4110 (2009).

30. S.T. Oyama, D. Lee, P. Hacarlioglu, R.F. Saraf, Theory of hydrogen permeability in nonporous silica membranes, *J. Membr. Sci.* 244, 45–53 (2004).

31. J.S. Bunch, S.S. Verbridge, J.S. Alden, A.M. Van der Zande, J.M. Parpia, H.G. Craighead, P.L. McEuen, Impermeable atomic membranes from graphene sheets, *Nano Lett.* 8, 2458–2462 (2008).

32. M.D. Fischbein, M. Drndic, Electron beam nanosculpting of suspended graphene sheets, *Appl. Phys. Lett.* 93, 113107 (2008).

33. P. Kuhn, A. Forget, D.S. Su, A. Thomas, M. Antonietti, From microporous regular frameworks to mesoporous materials with ultrahigh surface area: Dynamic reorganization of porous polymer networks, *J. Am. Chem. Soc.* 130, 13333–13337 (2008).

34. K. Sint, B. Wang, P. Kral, Selective ion passage through graphene nanopores, *J. Am. Chem. Soc.* 130, 16448–16449 (2008).

35. T. Thonhauser, V.R. Cooper, S. Li, A. Puzder, P. Hyldgaard, D.C. Langreth, van der Waals density functional: Self-consistent potential and the nature of van der Waals bond, *Phys. Rev. B* 76, 125112 (2007).

36. M. Dion, H. Rydberg, E. Schroder, D.C. Langreth, B.I. Lundqvist, Van der Waals density functional for general geometries, *Phys. Rev. Lett.* 92, 246401 (2004).

37. D.C. Langreth, B.I. Lundqvist, S.D. Chakarova-Kack, V.R. Cooper, M. Dion, P. Hyldgaard, A. Kelkkanen, J. Kleis, L.Z. Kong, S. Li, P.G. Moses, E. Murray, A. Puzder, H. Rydberg, E. Schroder, T. Thonhauser, A density functional for sparse matter, *J. Phys.: Condens. Matter* 21, 084203 (2009).

38. S.D. Chakarova-Kack, Q. Borck, E. Schroder, B.I. Lundqvist, Adsorption of phenol on graphite(001) and α-Al_2O_3: Nature of van der Waals bonds from first principle calculation, *Phys. Rev. B* 74, 155402 (2006).

39. S.D. Chakarova-Kack, E. Schroder, B.I. Lundqvist, D.C. Langreth, Application of van der Waals density functional to an extended system: Adsorption of benzene and naphthalene on graphite, *Phys. Rev. Lett.* 96, 146107 (2006).

40. L. Kong, V.R. Cooper, N. Nijem, K. Li, J. Li, Y.J. Chabal, D.C. Langreth, Theoretical and experimental analysis of H_2 binding in a prototypical metal-organic framework material, *Phys. Rev. B* 79, 081407(R) (2009).

41. X. Gonze, J.-M. Beuken, R. Caracas, F. Detraux, M. Fuchs, G.M. Rignanese, L. Sindic, M. Verstraete, G. Zerah, F. Jollet, M. Torrent, A. Roy, M. Mikami, P. Ghosez, J.-Y. Raty, D.C. Allan, First principles computation of material properties: The ABNIT software project, *Comput. Mater. Sci.* 25, 478–492 (2002).

42. G. Kresse, J. Furthmuller, Efficient iterative schemes for ab initio total-energy calculations using a plane wave basis set, *Phys. Rev. B* 54, 11169–11186 (1996).

43. G. Kresse, J. Furthmuller, Efficiency of ab-initio total energy calculations for metals and semiconductor using a plane wave basis set, *Comput. Mater. Sci.* 6, 15–50 (1996).

44. P.E. Blochl, Projector augmented-wave method, *Phys. Rev. B* 50, 17953–17979 (1994).

45. G. Kresse, D. Joubert, From ultrasoft pseudopotentials to the projector augumented wave method, *Phys. Rev. B* 59, 1758–1775 (1999).

46. X.R. Wang, X.L. Li, L. Zhang, Y. Yoon, P.K. Weber, H.L. Wang, J. Guo, H.J. Dai, N-doping of graphene through electrothermal reactions with ammonia, *Science* 324, 768–771 (2009).

47. A. Puzder, M. Dion, D.C. Langreth, Binding energy in benzene dimmers: Nonlocal density functional calculations, *J. Chem. Phys.* 124, 164105 (2006).

Chapter 5

Graphene-Based Materials in Biosensing and Energy Storage Applications

5.1 Electrochemical Biosensors Based on Graphene

Graphene was found to exhibit excellent electrochemical behavior, thereby indicating that it is a promising electrode material for electroanalysis [1,2]. To date, several reports confirmed graphene-based electrochemical sensors are suitable candidates for various bioanalysis and environmental analysis [3–6].

5.1.1 Graphene-Based Enzymatic Biosensors

Graphene was found to exhibit high electrocatalytic activity toward H_2O_2. Further, it serves as an excellent platform for the direct electrochemistry of glucose oxidase (GOD); therefore, graphene could be an excellent electrode material for oxidase biosensors. For example, there are numerous reports available in the literature about graphene-based glucose biosensors [3–9]. Shan et al. [3] reported the first graphene-based glucose biosensor with a graphene/polyethylenimine-functionalized ionic liquid nanocomposites modified electrode that exhibited a wide linear glucose response (2 to 14 mM) together with good reproducibility (relative standard deviation of the current response to 6 mM glucose at −0.5 V was 3.2% for 10 successive measurements) and high stability (response current +4.9% after 1 week) [3].

Glucose biosensors based on chemically reduced graphene oxide (CR-GO) have also been developed by Zhou et al. [10]. The biosensor exhibited substantially enhanced amperometric signals for sensing glucose with a wide linear range (0.01–10 mM) together with high sensitivity (20.21 μA mM cm^{-2}) and a low detection limit of 2.0 mM (S/N = 3). The linear range for glucose detection was wider than that on electrodes based on other carbon materials, such as carbon nanotubes (CNTs) [9] and carbon nanofibers [11]. The detection limit for glucose at the GOD/CR-GO/GC (glassy carbon) electrode (2.0 mM at −0.20 V) is lower than the reported carbon-based biosensors, such as CNT paste [12], CNT nanoelectrodes [13], carbon nanofibers [11], exfoliated graphite nanoplatelets [8], and highly ordered mesoporous carbon [14]. The response at the GOD/CR-GO/GC electrode to glucose was very fast (9 ± 1 s to attain a steady-state response) and highly stable (91% signal retention for 5 h), which makes the GOD/CR-GO/GC electrode a rapid and highly stable biosensor to continuously measure the plasma glucose level for the diagnosis of diabetes.

In another work, graphene dispersed on biocompatible chitosan was also employed to construct glucose biosensors [6]. In this work, it was evident that chitosan helped to form a well-dispersed graphene suspension and to immobilize the enzyme molecules, and the graphene-based enzyme sensor exhibited excellent sensitivity (37.93 mA mM^{-1} cm^{-2}) and long-term stability for measuring glucose.

Biosensors based on graphene/metal nanoparticles (NPs) have also been developed. Shan et al. [15] developed a biosensor based on graphene, gold NPs, and chitosan composites film that exhibited good electrocatalytical activity toward H_2O_2 and O_2. Wu et al. [7] designed a glucose biosensor based on GOD, graphene, platinum NPs, and chitosan with a detection limit of 0.6 mM glucose. These enhanced performances were attributed to the large surface area and good electrical conductivity of graphene and the synergistic effect of graphene and metal NPs [7,15].

Zhou et al. [16] reported an ethanol biosensor based on graphene-antidiuretic hormone (ADH). The ADH-graphene-GC electrode exhibited a faster response, wider linear range, and lower detection limit for ethanol detection compared with ADH-graphite/GC and ADH/GC electrodes. This observed enhanced performance can be explained by the effective transfer of substrate and products through graphene matrixes containing enzymes as well as the inherent biocompatibility of graphene [10].

5.1.2 Graphene-DNA Biosensors

The advantages of electrochemical DNA sensors include high sensitivity, high selectivity, and cost-effectiveness for the detection of selected DNA sequences or mutated genes associated with human disease and promise to provide a simple, accurate, and inexpensive platform for patient diagnosis [17,18]. Electrochemical DNA sensors further allow device miniaturization for samples with tiny volume [10]. Among various electrochemical DNA sensors, the sensor based on the direct oxidation of DNA is simple and robust [10,18].

Zhou et al. [10] reported an electrochemical DNA sensor based on graphene (CR-GO). As evident from Figure 5.1, the current responses of the four free bases of DNA—guanine (G), adenine (A), thymine (T), and cytosine (C)—on the CR-GO/GC electrode (Figure 5.1a) are all separated efficiently, indicating that

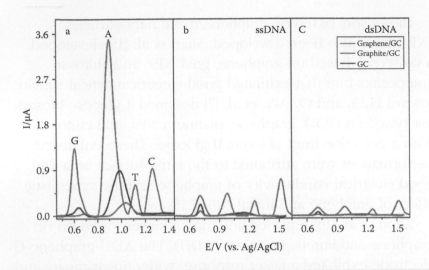

Figure 5.1 Differential pulse voltammetry (DPV) for (a) a mixture of DNA free base (G, A, T, and C), (b) ssDNA and (c) dsDNA in 0.1 *M* pH 7.0 phosphate-buffered saline (PBS) at graphene/GC, graphite/GC, and bare GC electrodes. Concentration of G, A, T, C, ssDNA, or dsDNA: 10 mg ml^{-1}. (Reproduced with permission from M. Zhou, Y.M. Zhai, S.J. Dong, *Anal. Chem.* 81, 5603–5613, 2009. Copyright 2009 Nature Publishing Group.)

CR-GO/GC can simultaneously detect four free bases, whereas this is not possible with graphite or glassy carbon. This was attributed to the antifouling properties and the high electron transfer kinetics for oxidation of bases on the CR-GO/GC electrode [10], which resulted from the high density of the edge-plane-like defects and oxygen-containing functional groups on CR-GO, which provided many active sites and promoted electron transfer between the electrode and species in solution [16,19,20]. From Figures 5.1b and 5.1c, it is evident that the CR-GO/GC electrode was capable of efficiently separating all four DNA bases in both single-stranded DNA (ssDNA) and double-stranded DNA (dsDNA), which are more difficult to oxidize than free bases, at physiological pH without the need of a prehydrolysis step. Further, this electrode provided a single-nucleotide polymorphism (SNP) site for short oligomers with a particular sequence without any hybridization or labeling processes [10]. This was attributed to the unique physicochemical properties of CR-GO (the single-sheet nature, high conductivity, large surface area, antifouling properties, high electron transfer kinetics, etc.) [10].

5.1.3 Graphene Sensors for Heavy Metal Ion Detection

Graphene-based electrochemical sensors also find potential application in environmental analysis for the detection of heavy metal ions (Pb^{2+} and Cd^{2+}) [21,22]. Li et al. [21,22] demonstrated that electrochemical sensors based on Nafion-graphene composite film not only exhibited improved sensitivity for the detection of Pb^{2+} and Cd^{2+}, but also alleviated the interferences due to the synergistic effect of graphene nanosheets and Nafion [21]. Further, the stripping current signal was greatly enhanced at the

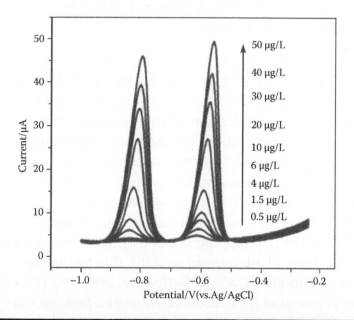

Figure 5.2 **Striping voltammograms for the different concentrations of Cd^{2+} and Pb^{2+} on an in situ plated Nafion-G-BFE (boron film electrode) in solution containing 0.4 mg L^{-1} Bi^{3+}. (Reproduced with permission from J. Li, S.J. Guo, Y.M. Zhai, E.K. Wang, *Anal. Chim. Acta* 649, 196–201, 2009.)**

graphene electrodes. It was evident from Figure 5.2 that the stripping current signal was well distinguished. The linear range for the detection of Pb^{2+} and Cd^{2+} was wide (0.5 to 50 mg L^{-1} and 1.5 to 30 mg L^{-1} for Pb^{2+} and Cd^{2+}, respectively). The detection limits ($S/N = 3$) were 0.02 mg L^{-1} for both Cd^{2+} and Pb^{2+}, which were more sensitive than that of Nafion film-modified bismuth electrode [23] and ordered mesoporous carbon-coated glassy carbon electrode (GCE) [24] and comparable to bismuth film electrode coated with Nafion/CNT [25]. The enhanced performance was attributed to the unique properties of the graphene (nanosize graphene sheet, nanoscale thickness of these sheets, and high conductivity), which endowed the capability to strongly adsorb target ions, enhanced the surface concentration, improved the sensitivity, and alleviated the fouling effect of surfactants [21,22].

5.1.4 Graphene for the Rapid Sequencing of DNA Molecules

To date, one of the greatest challenges in biotechnology is to establish the base sequence of individual DNA molecules without the need for polymerase chain reaction (PCR) amplification or other modification of the molecule [26]. Work by Sanger et al. [27] proved the necessity of sequencing of the human genome [28,29]. This was accomplished by shotgun sequencing, that is, breaking the sample into small random fragments and amplifying them, sequencing these fragments using the Sanger method, and merging these sequences by determining the overlapping areas by their base sequence. There are many challenges in making this technology more accurate, comprehensive, cost-effective, and fast. DNA amplification is essential in a resource-intensive process that can also introduce errors. In addition, repetitive regions larger than the Sanger read length are extremely difficult to sequence.

Numerous improvements to optimize various aspects of the sequencing process are in the pipeline [30,31]. Further, single-molecule sequencing techniques are being developed that present an alternative to the Sanger method [26], and they offer an increase in speed, reduction in cost, reduction in error rate, and increase in read length [26].

The potential for single-molecule sequencing with biological nanopores was first explored with naturally occurring R-hemolysin (RHL) proteins, which spontaneously embed themselves in a lipid bilayer and form a nanopore [32]. This RHL pore was studied using electrophysiology, in which a patch-clamp amplifier records the ion current through the protein pore while a DNA molecule translocates through it under the influence of an applied transmembrane electric field acting on the negatively charged backbone [33–36]. Both ssDNA and dsDNA have been studied. ssDNA can translocate through a pore as small as 1.5 nm [37]; 3 nm is large enough for dsDNA [38]. The lower limit for ssDNA translocation can be expected to be the single-nucleotide size of $wNT \approx 1$ nm [26]. When different blocking currents for different homomers were first recorded, it was suggested that this would allow for rapid sequencing of individual DNA molecules while they translocate through the pore [39,40]. However, the nanopore is too deep to realize single-base resolution, and the current signal is too small to allow for rapid readout of the current. Both may be addressed using exonuclease to break up the DNA and modifying the nanopore to slow the translocation [41]. However, the individual bases enter the nanopore in a different order than they are in the DNA. In addition, biological nanopores and the lipid bilayer membrane in which they are embedded are only stable within a small range of temperature, pH, chemical environments, and applied electric fields, limiting practical applications. On the other hand, solid-state nanopores are free from these limits [42]. They have been fabricated in Si_3N_4 membranes [43], SiO_2 membranes [44], and polymer films [45].

Transverse-conductance-based sequencing is a particularly promising next-generation sequencing technology [46–50].

The idea behind the technology is that different bases have different local electronic densities of states with different spatial extent owing to their different chemical composition. This idea is actively being tested using the conducting tip of a scanning tunneling microscope (STM) over immobilized DNA on a substrate [51–53]. If, instead, the bases are passing one by one through a voltage-biased tunnel gap inside a solid-state nano-pore, they will alternately change the current based on how the localized base states contribute to the tunnel current. However, making nanoelectrodes that are sufficiently thin to resolve the DNA conductance with single-base resolution as well as getting the nanoelectrodes aligned with the nanopore is challenging and has not been realized yet. Recently, Postma [26] proposed graphene nanogaps for DNA sequencing, using graphene as the electrode as well as the membrane material (Figure 5.3).

Postma [26] claimed that graphene is an ideal material for sequencing due to (1) its single-atom thickness, which enables

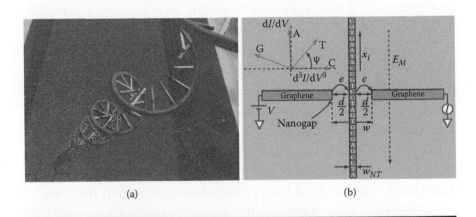

(a)   (b)

Figure 5.3 (See color insert.) (a) The individual bases of the ssDNA molecule (backbone in green, bases in alternating colors) sequentially occupy a gap in graphene (hexagonal lattice) while translocating through it. Their conductance is read, revealing the sequence of the molecule. The contacting electrodes to the graphene nanogap (gold, yellow) are on the far left and right sides of this image. (b) Schematic representation of the transverse conductance technique. (Reproduced with permission from H.W. Ch. Postma, *Nano. Lett.* 10, 420–425, 2010.)

transverse conductance measurements with single-base resolution; (2) its ability to survive large transmembrane pressures [54–56]; and (3) its intrinsic conducting properties. The last property is especially advantageous because the membrane is the electrode, automatically solving the problem of having to fabricate nanoelectrodes that are exactly aligned with a nanogap. Alternatively, various methods can be used to obtain graphene nanogaps. They may be fabricated by nano-lithography with an STM [57] in a method similar to that used for cutting CNTs [58]. Other possible fabrication methods are electromigration [59], local anodic oxidation [60], transmission electron microscopic (TEM) nanofabrication [61], or catalytic nanocutting [62,63].

The ideal nanogap width is 1.0 to 1.5 nm, allowing the ssDNA to pass through it in an unfolded state [37] as well as ensuring a large transverse current. To estimate the expected translocation speed and therefore the sequencing rate, the graphene nanogap can be compared to both solid-state and biological nanopores. The DNA translocation speed is typically much larger in solid-state nanopores than in biological nanopores owing to their large difference in size, aspect ratio, and DNA-pore interaction strength [64–69]. For pore sizes that are small compared to the ssDNA width, the bases stick to the side of the nanogap, lagging behind the backbone, while the molecule moves through the gap [70]. The translocation speed is also influenced by the unfolding speed of the DNA into the pore [38,68]. The RHL pore geometry is similar to the graphene nanogap proposed here since (1) the ideal nanogap width of 1.0–1.5 nm is similar and (2) the narrowest region of the RHL pore has a similar thickness to the graphene sheet. This may result in similar DNA-graphene nanogap interaction strength.

An advantage of graphene nanogaps is that their local atomic configuration can be imaged directly with the STM after the gap has been fabricated, allowing for a comprehensive comparison of measurements with theoretical calculations; Postma [26] reported a value of 3.6 µs per nucleotide as an average translocation time without any pre- or postprocessing. This value is well within range of the technique proposed by Postma [26].

To demonstrate the single-base resolution of the proposed graphene nanogap sequencing technique, Postma [26] proposed a numerical simulation based on the first-principles results of He et al. [50]. The results presented in Figure 5.4a showed a series of current peaks that corresponded to the individual nucleotides. It demonstrated that the individual bases can be resolved with this technique.

To demonstrate how a change in nanogap width affected the current, the simulation was performed for different nanogap widths $w = d + w\text{NT}$, where $w\text{NT} \approx 1$ nm is the single-nucleotide size (Figure 5.4). As the nanogap became wider, the current peaks became broader. In addition to the current peaks becoming wider, the overall current decreased exponentially with the nanogap width. As a result, a technique is required that will distinguish current changes due to base variations from changes due to nanogap width variations. To overcome this issue, Postma [26] proposed to use the nonlinear current-voltage characteristic to allow for base characterization independent of the nanogap width.

When the bases are aligned with the nanogap, the angle becomes stable, and its value is approximately the same for all nanogap widths. It is then used to determine the base type, as plotted in Figure 5.5a. The blue triangles indicate the actual base type; the red triangle indicates what the deduced base type is based on ψ. The histogram of recorded angles is presented in Figure 5.4a and shows four well-separated peaks due to the different base types. It is clear that this method can be used to sequence an individual DNA molecule, although nanogap width variations caused the current to vary by more than five orders of magnitude. When the nanogap was equal to 1.7 nm, the peaks became so broad that currents due to adjacent bases started to influence the current due to the base in the center of the nanogap. This led to a misidentification

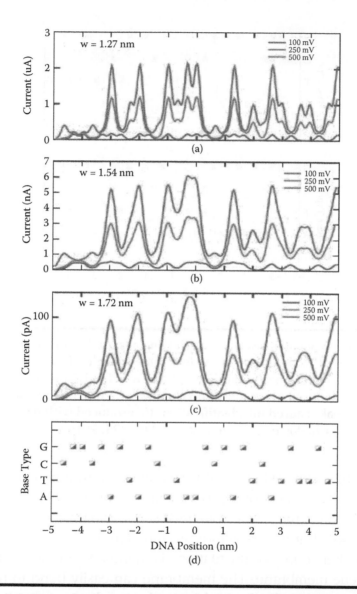

Figure 5.4 (See color insert.) (a)–(c) Current across the graphene nanogap while a single-stranded DNA molecule translocates through it for three different gap widths *w* and bias voltage levels as indicated. (d) For this simulation, a random sequence of CGG CGA GTA GCA TAA GCG AGT CAT GTT GT was used. (Reproduced with permission from H.W. Ch. Postma, *Nano. Lett.* 10, 420–425, 2010.)

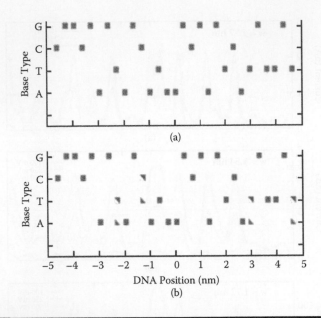

(a)

(b)

Figure 5.5 (See color insert.) **(a) The angle is used to determine the base type as described in the text (blue triangles, base type; red triangles, deduced base type from ψ). For *w* = 1.1–1.6 nm, the base type was deduced accurately. (b) When *w* = 1.7 nm, the overlapping current peaks caused misidentification. (Reproduced with permission from H.W. Ch. Postma, *Nano. Lett.* 10, 420–425, 2010.)**

of the base type (Figure 5.5b). This misidentification can be remedied by deconvolving the recorded current. This broadening is the prime source of sequencing errors, and the rate at which it occurs is indicated in Figure 5.6.

Another source of fluctuations is thermal vibrations of the graphene membrane. The membrane can easily bend in the direction perpendicular to the membrane plane owing to its single-atom thickness. Thermal vibrations of this bending mode [36] lead to a stochastic variation of the position of the nanogap with respect to the DNA longitudinal axis, and it limits the longitudinal resolution with which the base's transverse conductance can be measured. From recent studies, for few-sheet graphene membranes, the thermal noise amplitude can be estimated as 0.16 nm for a 0.6-nm thick and 500-nm long

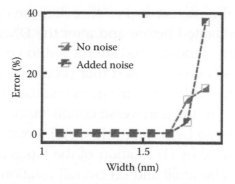

Figure 5.6 Sequence error rate with and without added Johnson-Nyquist noise. (Reproduced with permission from H.W. Ch. Postma, *Nano. Lett.* **10, 420–425, 2010.)**

membrane [36]. This was smaller than the 0.3-nm distance between bases, which means the single resolution will be possible despite these mechanical vibrations. Brownian motion of the ssDNA molecule will lead to a stochastic variation of the position of the nucleotides inside the nanogap. An upper limit to the magnitude of this effect can be estimated from the free diffusion of DNA when it is not inside the nanogap. The upper limit was 2 nm, which was larger than the base-to-base distance. However, the diffusion coefficient was most likely much smaller in the confined nanogap geometry [64]. In addition, it may be made smaller by functionalizing the nanogap [41]. Finally, driving the DNA through the nanogap more quickly will reduce τ, and consequently δx_0, even further.

To prevent a parasitic current pathway from the graphene surface through the ionic solvent that bypasses the nanogap, the graphene may be covered with a self-assembled mono-layer. According to Postma [26], this will also improve the wetting properties of the graphene surface, limit electrochemical reactions of the graphene surface in contact with the solvent, and prevent adhesion of the DNA to the graphene surface. Residual parasitic current between the unpassivated carbon atoms at the edge of the nanogap will cause an extra contribution to the current and its first derivate dI/dV. This offset can be calibrated before and after the DNA has translocated through the nanogap and subtracted to compensate for this effect. It has been calculated that geometric fluctuations of the nucleotides while they are in the nanogap can lead to large fluctuations in the transverse conductance, limiting how well nucleotides can be distinguished [48]. Rotational fluctuations can be caused by (1) rotation of the bases around the bond to the backbone $\delta\theta$ and (2) overall rotation of the DNA molecule inside the nanogap θ. Because the persistence length of the DNA molecule is much larger than the base-to-base distance, it can be expected that the second effect will lead to a similar fractional change in conductance for consecutive bases as the molecule translocates through the nanogap [48]. The presence of this effect and how it changes the current can therefore be deduced from a comparison of ψ for consecutive bases. The first effect will lead to a stochastic variation of the angle $\delta\theta$ around the average value θ.

A detailed study of how this changes the *nonlinear* conductance and ψ is the subject of future research. These conductance fluctuations may be reduced by stabilizing the nucleotides while they are in the nanogap, for example, by functionalizing the nanogap with cytosine [50] or by the applied bias voltage [72]. As an alternative to the transverse conductance technique proposed by Postma [26], the presented device can also be used to directly detect voltage fluctuations due to the local and unique dipole moments of the bases [73,74].

5.2 Graphene for Energy Storage Applications

5.2.1 Transparent Electrodes Based on Graphene

Solution processing of chemically derived graphene and the depositions achieved soon led researchers to consider using the material in transparent conductors [75]. The demand for such coatings has grown rapidly due to optoelectronic devices, including displays, light-emitting diodes (LEDs), and solar cells [75]. While the current industry standard is indium tin oxide (ITO), CNTs have long been touted as a possible alternative due to their low dimensionality and ability to form a percolating conductive network at extremely low densities [75]. The same merits make graphene an obvious choice.

Mullen and coworkers demonstrated the first graphene-based transparent conductor [76]. Films were deposited by dip coating using GO followed by reducing them by thermal annealing. Sheet resistances as low as 0.9 kΩ were obtained at 70% transmittance [76]. While the performance was considerably less than that of ITO at 90% transmittance), the films were low cost and negated vacuum sputtering [76]. The group also used the film as the anode in a dye-sensitized solar cell, which had a power conversion efficiency (PCE) of 0.26%. Following this, a polymer solar cell with a PCE of 0.1% using a similar film was designed by Eda et al. [77,78]. The performance of these cells was less than that of the corresponding control devices on ITO, but they provided a proof of concept for low-cost transparent coatings based on graphene.

5.2.2 Ultracapacitors Based on Graphene

The surface area of a single graphene sheet is 2630 m^2/g, which is manyfold higher than the values derived from Brunauer-Emmett-Teller (BET) surface area measurements of activated carbons employed in current electrochemical double-layer capacitors. Ruoff et al. [79] pioneered chemically modified graphene (CMG) that were made from sheets of carbon one atom thick, functionalized as needed, and demonstrated in an ultracapacitor cell their performance. In addition, they have also measured the specific capacitances of graphene in aqueous and organic electrolytes as 135 and 99 F/g, respectively [79]. Also, high electrical conductivity gives these materials consistently good performance over a wide range of voltage scan rates. Based on these results, Ruoff et al. [79] confirmed that graphene is undoubtedly an exciting material for high-performance electrical energy storage devices.

Ultracapacitors based on electrochemical double-layer capacitance (EDLC) are electrical energy storage devices that store and release energy by nanoscopic charge separation at the electrochemical interface between an electrode and an electrolyte [79]. The energy stored is inversely proportional to the thickness of the double layer; as a result, these capacitors have an extremely high energy density compared to conventional dielectric capacitors [79]. They are able to store an enormous amount of charge, which can be delivered at huge power ratings than rechargeable batteries [79]. Ultracapacitors can be employed for a wide range of energy capture and storage applications. Further, they can be used either by themselves as the primary power source or in combination with batteries or fuel cells [79].

Some advantages of ultracapacitors over more traditional energy storage devices include high power capability, long life, a wide thermal operating range, lower weight, flexible packaging, and user friendliness [80]. Ultracapacitors are ideal for any application having a short load cycle and high-reliability requirement, such as energy recapture sources, including load cranes, forklifts, and electric vehicles [79–81]. Other applications that exploit an ultracapacitor's ability to nearly instantaneously absorb and release power include power leveling for electric utilities and factory power backup. A group of ultracapacitors can bridge the short time duration between a power failure and the startup of backup power generators.

Although the energy density of ultracapacitors is several-fold higher than conventional dielectric capacitors, it is still significantly lower than batteries or fuel cells. Coupling with batteries (or another power source) is still required for supplying energy for longer periods of time. Thus, there is a growing interest, for example, as enunciated by the U.S. Department of Energy, for increasing the energy density of ultracapacitors to be closer to the energy density of batteries [82].

In addition to the EDLCs described, another class of ultracapacitor that is based on pseudocapacitance can be employed. While the charge storage mechanism of EDLCs is nonfaradic, pseudocapacitance is based on faradic, redox reactions using electrode materials such as electrically conducting polymers and metal oxides. The energy densities of pseudocapacitance-based devices can be greater than EDLCs; however, the phase changes within the electrode due to the faradic reaction limits their lifetime and power density. The results reported here are based on EDLC ultracapacitor cells with CMG-based carbon electrode material [79].

Briefly, an ultracapacitor unit cell consists of two porous carbon electrodes that are isolated from electrical contact by a porous separator [83]. Current collectors of metal foil or carbon-impregnated polymers are used to conduct electrical current from each electrode. The separator and the electrodes are impregnated with an electrolyte, which allows ionic current to flow between the electrodes while preventing electronic current from discharging the cell [79]. A packaged ultracapacitor module, depending on the desired size and voltage, is constructed of multiple repeating unit cells [79]. The CMG system of individual sheets does not depend on the distribution of pores in a solid support to give it its large surface area; rather, every CMG sheet can move physically to adjust to the different types of electrolytes (their sizes, their spatial distribution). Thus, access to the very high surface area of CMG materials by the electrolyte can be maintained while preserving the overall high electrical conductivity for such a network [84,85].

Due to the relatively high electrical resistance of the activated carbon materials, commercial electrodes are limited in thickness and usually contain conductive but low-surface-area additives such as carbon black to enable rapid electrical charge transfer from the cell [79]. The measured conductivity of these CMG materials ($\sim 2 \times 10^2$ S/m) is close to that of pristine graphite. The high electrical conductivity of the graphene materials eliminates the need for conductive fillers and allows increased electrode thickness. Increasing the electrode thickness and eliminating additives leads to an improved ratio of electrode material to collector/separator, which in turn further increases the energy density of the packaged ultracapacitor.

CMG materials can be synthesized with several methods into various morphologies [86], and they can be kept suspended in solution [87,88] configured into paperlike materials [89,90] and incorporated into polymer [91] or glass/ceramic [92] nanocomposites. Ruoff et al. [79] employed a special form of CMG that could be readily incorporated into ultracapacitor test cell electrodes in this study. The special form of CMG was synthesized by suspending graphene oxide sheets in water followed by reduction in hydrazine hydrate. During reduction, the individual graphene sheets agglomerated into approximately particles with a diameter of 15–25 μm and were measured using scanning electron microscopy (SEM). Furthermore, Ruoff et al. [79] determined the C/O and C/N atomic ratios using elemental analysis by combustion of the CMG powder sample.

Figure 5.7a shows an SEM image of the surface of the CMG agglomerate particles. Figure 5.7b is a TEM image that shows individual graphene sheets extending from the outer surface. From the images, Ruoff et al. [79] showed how both sides of the individual sheets at the surface of the agglomerate were exposed to the electrolyte. Ruoff et al. [79] further evaluated the surface area of the CMG agglomerate as 705 m^2/g by the N_2 absorption BET method. Furthermore, the graphene sheets located within the agglomerated particles may not be accessible by the electrolyte; however, sheets at the surface can provide an indication of CMG's potential for use in ultracapacitor electrodes.

Electrode material characterization can be performed using either a two- or a three-electrode configuration. However, routinely the two-electrode test cell configuration was used because it provides the most accurate measure of a material's performance for electrochemical capacitors [93]. The CMG particles were formed into electrodes using a polytetrafluoroethylene

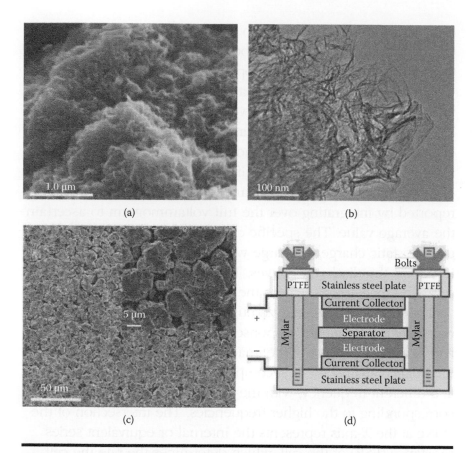

Figure 5.7 (a) SEM image of CMG particle surface. (b) TEM image showing individual graphene sheets extending from CMG particle surface (c), low and high (inset) magnification SEM images of CMG particle electrode surface, and (d) schematic of test cell assembly. (Reproduced with permission from M.D. Stoller, S. Park, Y. Zhu, J. An, R.S. Ruoff, *Nano Lett.* **8**, 3498–3502, 2008.)

(PTFE) binder, and a stainless steel test fixture was used for electrical testing of the assembled cell. Figure 5.7c shows two SEM images (low and high magnification) of the surface of an as-prepared electrode. Figure 5.7d shows a schematic of the two-electrode ultracapacitor test cell and fixture assembly employed in the work of Stoller et al. [79]. CMG-based ultracapacitor cells were then tested with three different electrolytes commonly used in commercial EDLCs. The electrolytes consisted of an aqueous electrolyte (5.5 *M* KOH) and two organic electrolyte systems: tetraethylammonium tetrafluoroborate (TEABF$_4$) in acetonitrile (AN) solvent and TEABF$_4$ in propylene carbonate (PC) solvent.

Ruoff et al. [79] further evaluated the performance of the ultracapacitor cells using cyclic voltammetry (CV), electrical impedance spectroscopy (EIS), and galvanostatic charge/discharge. The voltammograms and galvanostatic charge/discharge were used to calculate the specific capacitance of the CMG electrodes. The specific capacitance using the CV curves was reported by integrating over the full voltammogram to ascertain the average value. The specific capacitance determined from galvanostatic charge/discharge was calculated from the slope (dV/dt) of the discharge curves. Table 5.1 shows the results of specific capacitance (F/g) for the two methods.

The EIS data were analyzed using Nyquist plots. Nyquist plots represent the frequency response of the CMG electrode/electrolyte system and a plot of the imaginary component Z'' of the impedance against the real component Z'. Each data point was at a different frequency, with the lower left portion of the curve corresponding to the higher frequencies. The intersection of the curve at the X-axis represents the internal or equivalent series resistance (ESR) of the cell, which determines the rate the cell can be charged/discharged (power capability). The slope of the

Table 5.1   Specific Capacitance (F/g) of CMG Material

	Galvanostatic Discharge (mA)		Scan Rate (mV/s)	
Electrolyte	10	20	20	40
KOH	134	128	100	107
TEABF$_4$/PC	94	91	82	80
TEABF$_4$/AN	99	95	99	85

Source: Reproduced with permission from M.D. Stoller, S. Park, Y. Zhu, J. An, R.S. Ruoff, *Nano Lett.* 8, 3498–3502, 2008.

45° portion of the curve is called the Warburg resistance and is a result of the frequency dependence of ion diffusion/transport in the electrolyte. However, Ruoff et al. [79] chose the morphology of the CMG material such that only a portion of the graphene sheets (those at the surface of the particles) were exposed to electrolyte. The specific capacitances of the CMG were on the order of 100 F/g, indicating that the graphene material worked well with current commercial electrolytes, had good electrical conductivity, and had promising charge storage capability.

The cyclic voltammograms (shown in Figure 5.8) obtained were nearly rectangular in shape, indicating good charge

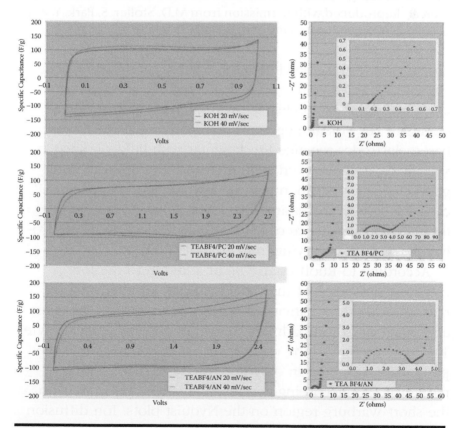

Figure 5.8 CV (left) and Nyquist (right) plots of CMG material with KOH electrolyte (top), TEABF$_4$ in propylene carbonate (middle), and TEABF$_4$ in acetonitrile (bottom). (Reproduced with permission from M.D. Stoller, S. Park, Y. Zhu, J. An, R.S. Ruoff, *Nano Lett.* **8**, 3498–3502, 2008.)

Table 5.2 Specific Capacitance (F/g) of CMG Material in KOH Electrolyte with Scan Rate

Scan Rate (mV/s)	Average Specific Capacitance (F/g)
20	101
40	106
100	102
200	101
300	96
400	97

Source: Reproduced with permission from M.D. Stoller, S. Park, Y. Zhu, J. An, R.S. Ruoff, *Nano Lett.* 8, 3498–3502, 2008.

propagation within the electrodes. For activated-carbon-based electrodes, the voltammogram shape and the specific capacitance can significantly degrade with increase in the voltage scan [94]. The voltammograms corresponding to the CMG-based electrodes remained rectangular with little variance in specific capacitance even at 40 mV/s. Another indication of good charge propagation was the low variation of specific capacitance with increasing scan rates. Table 5.2 summarizes the specific capacitance of CMG electrodes with KOH electrolyte for scan rates varying from 20 to 400 mV/s.

According to Ruoff et al. [79], in addition to the low electrical resistance of the CMG material, another possible reason for the insensitivity to varying scan rates is the short and equal diffusion path length of the ions in the electrolyte. This may be due to the electrolyte not penetrating into the particulate, resulting in only the graphene sheets at the particulate surface being accessed. This could also explain the short Warburg region on the Nyquist plots. Ion diffusion into the interior of the agglomerate would result in greater variations in ion diffusion path lengths and an increased obstruction in the ion movement, resulting in a much larger Warburg region.

The high conductivity of the CMG material also contributes to the low ESR of the cells. Ruoff et al. [79] measured the internal cell resistance (real Z' axis from Nyquist plots) to be 0.15 Ω (24 kHz), 0.64 Ω (810 kHz), and 0.65 Ω (500 kHz) for KOH, TEABF$_4$/PC, and TEABF$_4$/AN electrolytes, respectively. Further, the change in specific capacitance with respect to voltage also remained relatively linear at the higher voltages. Test cells with KOH electrolyte were cycled to 1 V, cells with PC were cycled to 2.7 V, and those with AN were cycled to 2.5 V. The presence of a low percentage of functional groups in the CMG material may contribute to a small amount of pseudocapacitance; however, the relatively linear increase in current with increasing voltage indicated that the charge was primarily nonfaradic in nature [95]. The specific capacitance of KOH at 40 mV/s (scan rate) during charging remained almost a constant 116 F/g between 0.1 and 0.9 V. The specific capacitance of the organic electrolyte AN during discharge at 20 mV/s was 100 F/g in the range from 1.5 V to fully discharged at 0 V. The specific capacitance of the organic electrolyte using PC during discharge at 20 mV/s was about 95 F/g in the range from 2.0 V to fully discharged at 0 V.

Thus, Ruoff et al. [79] confirmed that CMGs with good electrical conductivity and very large (and in principle completely accessible) surface areas are extremely promising candidates for EDLC ultracapacitors. In addition, these CMG materials are based on abundantly available and cost-effective graphite. Ultracapacitors based on these materials could have the cost and performance that would dramatically accelerate their adoption in a wide range of energy storage applications.

5.2.3 N-Doped Graphene for Oxygen Reduction in Fuel Cells

Platinum NPs were considered as the best catalyst for the oxygen reduction reaction (ORR) in fuel cells for decades, although the platinum-based electrode suffers from two major shortcomings: susceptibility to time-dependent drift and CO deactivation [96–98]. Furthermore, the high cost of the platinum catalysts together with the limited reserves of platinum in nature have been major issues for the fuel cell market for commercial applications. As a result, the large-scale practical application of fuel cells has not been realized, although alkaline fuel cells with platinum as an ORR electrocatalyst were developed for the Apollo lunar mission in the 1960s [99].

There are several research groups working intensively to minimize or to replace platinum-based electrodes in fuel cells [100–103]. Qu et al. [104] studied the possibility of employing nitrogen-doped graphene sheets (N-doped graphene) as a catalyst for the ORR process in fuel cells. In their study, Qu et al. [104] investigated the CVD-deposited N-graphene sheets for ORR at the cathode in alkaline fuel cells. The results of Qu et al. [104] indicated that the N-doped graphene showed a much better electrocatalytic activity, long-term stability, and tolerance to crossover and poison effects than the commercially available platinum-based electrodes (C2–20, 20% platinum on Vulcan XC-72R; E-TEK) for oxygen reduction.

A modified CVD process has been employed to synthesize N-doped graphene [105]. Briefly, a thin layer of nickel (300 nm) was deposited on a SiO_2/Si substrate by a sputtering technique. The nickel-coated SiO_2/Si wafer was then heated to 1000°C within a quartz tube furnace under a high-purity argon atmosphere. Following this, a nitrogen-containing reaction gas mixture (NH_3:CH_4:H_2:Ar = 10:50:65:200 standard cubic centimeters per minute) was introduced into the quartz tube and kept flowing for 5 min, followed by purging with a flow of NH_3 and argon for another 5 min. The sample was then rapidly moved from the furnace center (1000°C) under argon protection. The resultant N-doped graphene film can be readily etched off from the substrate by dissolving the residual nickel catalyst layer in an aqueous solution of HCl [105], allowing the freestanding N-doped graphene sheets to be transferred onto substrates suitable for subsequent investigation.

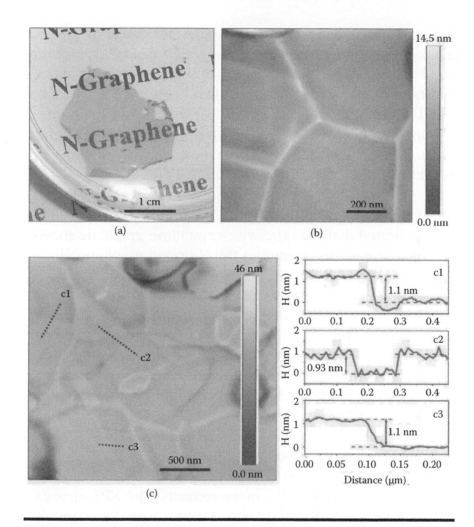

Figure 5.9 (a) A digital image of a transparent N-graphene film floating on water after removal of the nickel layer by dissolving in an aqueous acid solution. (b), (c) AFM images of the N-graphene film and the corresponding height analyses along the lines marked in the AFM image (c1 to c3 in panel c). (Reproduced with permission from L. Qu, Y. Liu, J.B. Baek, L. Dai, *ACS Nano* 4, 1321–1326, 2010.)

Figure 5.9a shows a freestanding film of about 4 cm² floating on water after removing the nickel layer by HCl. Similar to CVD-synthesized C-graphene films [105], the N-doped graphene film obtained in this method was flexible and transparent, consisting of only one- or a few-layer graphite sheets. The atomic force microscopic (AFM) image in Figure 5.9b shows a smooth surface with wrinkles due to its pliability. Further, the thickness of the layers was evaluated from the layered structure shown in Figure 5.9c; it was found to be in the range of 0.9 to 1.1 nm. The substrate-free N-graphene sheets can be directly transferred onto TEM grids for further characterization. Unlike most C-graphene films, the electron diffraction patterns [106,107] of the N-doped graphene film (inset of Figure 5.10a) shows a ringlike diffraction pattern with dispersed bright spots. The observed difference indicated that the otherwise-crystalline graphene sheets became partially disoriented in the N-doped graphene film due to the structure distortions caused by the intercalation of nitrogen atoms into its graphitic plans. The cross-sectional view of the suspended edge of the *as-synthesized* N-doped graphene film shows, again, only a few layers (typically, two to eight layers) of graphene sheets (Figures 5.10b, 5.10c, and 5.10d). Adjacent interlayer distances in the N-doped graphene film were measured as 0.3–0.4 nm close to the *d*-spacing of the (002) crystal plane (0.335 nm) of bulk graphite with slight distortion [108].

To ascertain the number of nitrogen atoms in the N-doped graphene structure, Qu et al. [104] performed X-ray photoelectron spectroscopic (XPS) measurements. The XPS survey spectrum for the N-doped graphene shows a predominant narrow graphitic C1s peak at 284.2 eV [109,110] along with an N1s peak at about 400 eV, as shown in Figure 5.11. On the other hand, the XPS survey spectra for both the N-doped graphene and C-graphene over a wide range of binding energies (0–1000 eV) exhibited an O1s peak at about 540 eV [111,112], possibly due to the incorporation of physically adsorbed oxygen as in the case with CNTs [113]. Thus, the higher O1s peak relative to the corresponding C1s peak seen for the N-doped graphene than that of the C-graphene suggests stronger O_2 adsorption on the former, which is an additional

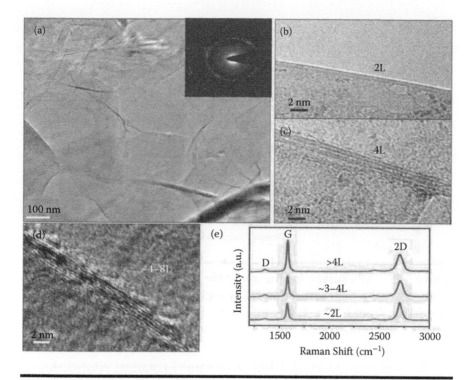

Figure 5.10   TEM and Raman analyses of the N-graphene films: (a) Low-magnification TEM image showing a few layers of the CVD-grown N-graphene film on a grid. Inset shows the corresponding electron diffraction pattern. (b) to (d) High-magnification TEM images showing edges of the N-graphene film regions consisting of (b) two, (c) four, and (d) about four to eight graphene layers. (e) The corresponding Raman spectra of the N-graphene films of different graphene layers on an SiO$_2$/Si substrate. (Reproduced with permission from L. Qu, Y. Liu, J.B. Baek, L. Dai, *ACS Nano* 4, 1321–1326, 2010.)

advantage as the ORR electrode. The absence of any nickel peak in the XPS spectrum for the N-doped graphene clearly indicates that the nickel catalyst residues were completely removed by the HCl solution. Further, in the high-resolution XPS N1s spectrum, Qu et al. [104] observed the presence of both pyridine-like (398.3 eV) and pyrrolic (400.5 eV) nitrogen atoms within the graphene structure [114–116], and the calculated N/C atomic ratio was about 4 atomic%.

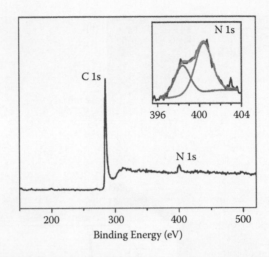

Figure 5.11   XPS survey for the *as-synthesized* N-graphene film. Inset shows the high-resolution N1s spectrum. (Reproduced with permission from L. Qu, Y. Liu, J.B. Baek, L. Dai, *ACS Nano* 4, 1321–1326, 2010.)

In addition, Qu et al. [104] performed rotating ring-disk electrode (RRDE) voltammograms to investigate the electrocatalytic activities of the N-doped graphene film for ORR in air-saturated 0.1 *M* KOH electrolyte. The C-graphene films prepared under similar CVD conditions, but without the introduction of NH_3 gas and commercially available platinum-loaded carbon supported by a GCE (Pt/C), were also investigated for comparison. Similar to the pure CNT electrode without N-doping, the C-graphene electrode showed a two step, two-electron process for oxygen reduction with the onset potential of about −0.45 and −0.7 V. Unlike the C-graphene electrode, the N-doped graphene electrode exhibited a one-step, four-electron pathway for the ORR, similar to N-doped CNTs. The steady-state catalytic current density at the N-doped graphene electrode was about three times higher than that of the Pt/C electrode over a large potential range (Figure 5.12a). The transferred electron number *n* per oxygen molecule at the N-doped graphene electrode was calculated by the Koutecky-Levich (K-L)

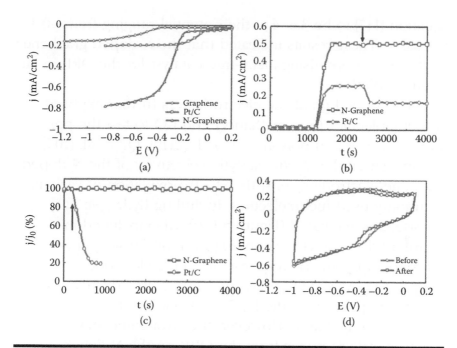

Figure 5.12 (a) RRDE voltammograms for the ORR in air-saturated 0.1 *M* KOH at the C-graphene electrode, Pt/C electrode, and N-graphene electrode. Electrode rotating rate: 1000 rpm. Scan rate: 0.01 V/s. Mass(graphene) = Mass(PtC) = Mass(N-graphene) = 7.5 µg. (b) Current density (*j*)-time (*t*) chronoamperometric responses obtained at the Pt/C (circle line) and N-graphene (square line) electrodes at −0.4 V in air-saturated 0.1 *M* KOH. The arrow indicates the addition of 2% (w/w) methanol into the air-saturated electrochemical cell. (c) Current (*j*)-time (*t*) chronoamperometric response of Pt/C (circle line) and N-graphene (square line) electrodes to CO. The arrow indicates the addition of 10% (v/v) CO into air-saturated 0.1 *M* KOH at −0.4 V; J_2 defines the initial current. (d) Cyclic voltammograms of N-graphene electrode in air-saturated 0.1 *M* KOH before (circle line) and after (square line) a continuous potentiodynamic sweep for 200,000 cycles at room temperature (25°C). Scan rate: 0.1 V/s. (Reproduced with permission from L. Qu, Y. Liu, J.B. Baek, L. Dai, *ACS Nano* 4, 1321–1326, 2010.)

equation [117] to be 3.6–4 at the potential ranging from −0.4 to −0.8 V. These results indicated that the N-doped graphene electrode is a promising metal-free catalyst for the ORR in an alkaline solution.

The N-doped graphene electrode was further exposed to fuel molecules (e.g., methanol) and CO to test the possible crossover and poison effects [118,119]. Qu et al. [104] also measured the electrocatalytic selectivity of the N-doped graphene electrode against the electro-oxidation of various commonly used fuel molecules, including hydrogen gas, glucose, and methanol (Figure 5.12b). Also included in Figure 5.12b is the corresponding current density j time t chronoamperometric response at a Pt/C electrode for comparison. As shown in Figure 5.12b, a 40% decrease in current appeared at the Pt/C electrode on the addition of 2% (w/w) methanol. However, the strong and stable amperometric response from the ORR on the N-doped graphene electrode remained unchanged after the addition of hydrogen gas, glucose, and methanol (Figure 5.12b). Such high selectivity of the N-doped graphene electrode toward the ORR and remarkably good tolerance to crossover effect can be attributed to the much lower ORR potential than that required for oxidation of the fuel molecules.

To examine the effect of CO on the electrocatalytic activity of the N-doped graphene electrode, 10% (v/v) CO in air was introduced into the electrolyte. The N-doped graphene electrode was insensitive to CO, whereas the Pt/C electrode was rapidly poisoned under the same conditions. Furthermore, Qu et al. [104] also performed continuous potential cycling to investigate the stability of the N-doped graphene electrode toward ORR. From Figure 5.12d, it is evident that no obvious decrease in current was observed after 200,000 continuous cycles between −1.0 and 0 V in air-saturated 0.1 M KOH.

References

1. R.L. McCreery, Advanced electrode materials for molecular electrochemistry, *Chem. Rev.* 108, 2646–2687 (2008).
2. J. Wang, Carbon nanotube based electrochemical biosensors: A review, *Electroanalysis* 17, 7–14 (2005).
3. C.S. Shan, H.F. Yang, J.F. Song, D.X. Han, A. Ivaska, L. Niu, Direct electrochemistry of glucose oxidase and biosensing for glucose based on graphene, *Anal. Chem.* 81, 2378–2382 (2009).
4. Z.J. Wang, X.Z. Zhou, J. Zhang, F. Boey, H. Zhang, Direct electrochemical reduction of single-layer graphene oxide and subsequent functionalization with glucose oxidase, *J. Phys. Chem. C* 113, 14071–14075 (2009).
5. H. Wu, J. Wang, X. Kang, C. Wang, D. Wang, J. Liu, et al., Glucose biosensor based on immobilization of glucose oxidase in platinum nanoparticles/graphene/chitosan nanocomposite film, *Talanta* 80, 403–406 (2009).
6. X.H. Kang, J. Wang, H. Wu, A.I. Aksay, J. Liu, Y.H. Lin, Glucose oxidase-graphene-chitosan modified electrode for direct electrochemistry and glucose sensing, *Biosens. Bioelectron.* 25, 901–905 (2009).
7. H. Wu, J. Wang, X.H. Kang, C.M. Wang, D.H. Wang, J. Liu, et al., Glucose biosensors based on immobilization of glucose oxidase in platinum nanoparticles/graphene/chitosan nanocomposite film, *Talanta* 80, 403–407 (2009).
8. J. Lu, L.T. Drzal, R.M. Worden, I. Lee, Simple fabrication of a highly sensitive glucose biosensor using enzymes immobilized in exfoliated graphite nanoplatelets Nafion membrane, *Chem. Mater.* 19, 6240–6246 (2007).
9. G.D. Liu, Y.H. Lin, Amperometric glucose biosensor based on self-assembling glucose oxidase on carbon nanotubes, *Electrochem. Commun.* 8, 251–256 (2006).
10. M. Zhou, Y.M. Zhai, S.J. Dong, Electrochemical biosensing based on reduced graphene oxide, *Anal. Chem.* 81, 5603–5613 (2009).
11. L. Wu, X.J. Zhang, H.X. Ju, Amperometric glucose sensor based on catalytic reduction of dissolved oxygen at soluble carbon nanofiber, *Biosens. Bioelectron.* 23, 479–484 (2007).
12. M.D. Rubianes, G.A. Rivas, Carbon nanotubes paste electrode, *Electrochem. Commun.* 5, 689–694 (2003).

13. Y.H. Lin, F. Lu, Y. Tu, Z.F. Ren, Glucose biosensors based on carbon nanotube nanoelectrode ensembles, *Nano Lett.* 4, 191–195 (2004).

14. M. Zhou, L. Shang, B.L. Li, L.J. Huang, S.J. Dong, Highly ordered mesoporous carbons as electrode material for the construction of electrochemical dehydrogenase and oxidase based biosensors, *Biosens. Bioelectron.* 24, 442–447 (2008).

15. C.S. Shan, H.F. Yang, D.X. Han, Q.X. Zhang, A. Ivaska, L. Niu, Graphene/AuNPs/chitosan nanocomposites film for glucose biosensing, *Biosens. Bioelectron.* 25, 1070 (2009).

16. M. Zhou, Y. M. Zhai, S. J. Dong, *Anal. Chem.* 2009, 81, 5603.

17. T.G. Drummond, M.G. Hill, J.K. Barton, Electrochemical DNA biosensors, *Nat. Biotechnol.* 21, 1192–1199 (2003).

18. C.E. Banks, T.J. Davies, G.G. Wildgoose, R.G. Compton, Electrocatalysis at graphite and carbon nanotube modified electrodes: Edge plane sites and tube ends are reactive sites, *Chem. Commun.* 829–841 (2005).

19. C.E. Banks, R.R. Moore, T.J. Davies, R.G. Compton, Investigation of modified basal plane pyrolytic graphite electrodes: Definitive evidence for the electrocatalytic properties of the ends of carbon nanotubes, *Chem. Commun.* 1804–1805 (2004).

20. C.E. Banks, R.G. Compton, Exploring the electrocatalytic sites of carbon nanotubes for NADH detection: An edge plane pyrolytic graphite electrode study, *Analyst* 130, 1232–1239 (2005).

21. J. Li, S.J. Guo, Y.M. Zhai, E.K. Wang, High-sensitivity determination of lead and cadmium based on the Nafion-graphene composite film, *Anal. Chim. Acta* 649, 196–201 (2009).

22. J. Li, S.J. Guo, Y.M. Zhai, E.K. Wang, Nafion-graphene nanocomposite film as enhanced sensing platform for ultrasensitive determination of cadmium, *Electrochem. Commun.* 11, 1085 (2009).

23. G. Kefala, A. Economou, A. Voulgaropoulos, A study of Nafion-coated bismuth-film electrodes for the determination of trace metals by anodic stripping voltammetry, *Analyst* 129, 1082–1090 (2004).

24. L.D. Zhu, C.Y. Tian, R.L. Yang, J.L. Zhai, Anodic stripping determination of lead in tap water at an ordered mesoporous carbon/nafion composite film electrode, *Electroanalysis* 20, 527–533 (2008).

25. H. Xu, L.P. Zeng, S.J. Xing, Y.Z. Xian, G.Y. Shi, Ultrasensitive voltammetric detection of trace lead (III) and cadmium (III) using MWCNTs-nafion/bismuth composite electrodes, *Electroanalysis* 20, 2655–2662 (2008).
26. H.W. Ch. Postma, Rapid sequencing of individual DNA molecules in graphene nanogaps, *Nano. Lett.* 10, 420–425 (2010).
27. F. Sanger, S. Nicklen, A.R. Coulson, DNA sequencing with chain-terminating inhibitors, *Proc. Nat. Acad. Sci. USA* 74, 5463–5467 (1977).
28. E. Lander, E.S. Lander, L.M. Linton, B. Birren, C. Nusbaum, M.C. Zody, et al., Initial sequencing and analysis of the human genome, *Nature* 409, 860–921 (2001).
29. J.C. Venter, M.D. Adams, E.W. Myers, P.W. Li, R.J. Mural, G.G. Sutton, et al., The sequence of human genome, *Science* 291, 1304–1351 (2001).
30. J. Shendure, R.D. Mitra, C. Varma, G.M. Church, Advanced sequencing technologies: Methods and goals, *Nat. Rev. Genet.* 5, 335–344 (2004).
31. C.P. Fredlake, D.G. Hert, E.R. Mardis, A.E. Barron, What is the future of electrophoresis in large-scale genome sequencing? *Electrophoresis* 27, 3689–3702 (2006).
32. J.J. Nakane, M. Akeson, A. Marziali, Nanopore sensors for nucleic acid assays, *J. Phys.: Condens. Matter* 15, R1365–R1393 (2003).
33. J. Henry, J. Chich, D. Goldschmidt, M. Thieffry, Blockade of a mitochondrial cationic channel by an addressing peptide: An electrophysiological study, *J. Membr. Biol.* 112, 139–147 (1989).
34. H.J. Bayley, Biotechnology applications in biomaterials, *J. Cell. Biochem.* 56, 177–182 (1994).
35. S.M. Bezrukov, I. Vodyanoy, V.A. Parsegian, Counting polymers moving through a single ion channel, *Nature* 370, 279–281 (1994).
36. J.O. Bustamante, H. Oberleithner, J.A. Hanover, A. Liepins, Patch clamp detection of transcription factor translocation along the nuclear pore complex channel, *J. Membr. Biol.* 146, 253–261 (1995).
37. X. Zhao, C.M. Payne, P.T. Cummings, J.W. Lee, Single strand DNA molecule translocation through nanopore gaps, *Nanotechnology* 18, 424018 (2007).

38. J.B. Heng, A. Aksimentiev, C. Ho, P. Marks, Y.V. Grinkova, S. Sligar, K. Schulten, G. Timp, The electromechanics of DNA in synthetic nanopore, *Biophys. J.* 90, 1098–1106 (2006).

39. M. Akeson, D. Branton, J.J. Kasianowicz, E. Brandin, D.W. Deamer, Microsecond time-scale discrimination among polycytidylic acid, polyadenylic acid and polyuridylic acid as homopolymers or as segments within single RNA molecules, *Biophys. J.* 77, 3227–3233 (1999).

40. A. Meller, L. Nivon, E. Brandin, J.A. Golovchenko, D. Branton, Rapid nanopore discrimination between single polynucleotide molecules, *Proc. Natl. Acad. Sci. USA* 97, 1079–1084 (2000).

41. J. Clarke, H. Wu, L. Jayasinghe, A. Patel, S. Reid, H. Bayley, Continuous base identification for single-molecule nanopore DNA sequencing, *Nat. Nanotechnol.* 4, 265–270 (2009).

42. C. Dekker, Solid state nanopores, *Nat. Nanotechnol.* 2, 209–215 (2007).

43. J. Li, D. Stein, C. McMullan, D. Branton, M.J. Aziz, J.A. Golovchenko, Ion beam sculpting at nanometer length scales, *Nature* 412, 166–169 (2001).

44. A.J. Storm, J.H. Chen, X.S. Ling, H.W. Zandbergen, C. Dekker, Fabrication of solid state nanopores with single nanometer precision, *Nat. Mater.* 2, 537–540 (2003).

45. A. Mara, Z. Siwy, C. Trautmann, J. Wan, F. Kamme, An asymmetric polymer nanopore for single molecule detection, *Nano Lett.* 4, 497–501 (2004).

46. M. Zwolak, M. Di Ventra, Electronic signature of DNA nucleotides via transverse transport, *Nano Lett.* 5, 421–424 (2005).

47. J. Lagerqvist, M. Zwolak, M. Di Ventra, Fast DNA sequencing via transverse electronic transport, *Nano Lett.* 6, 779–782 (2006).

48. X. Zhang, P.S. Krstic, R. Zikic, J.C. Wells, M. Fuentes-Cabrera, First-principles transversal DNA conductance deconstructed, *Biophys. J.* 91, L04–L06 (2006).

49. M. Zwolak, M. Di Ventra, Colloqium: Physical approaches to DNA sequencing and detection, *Rev. Mod. Phys.* 80, 141–165 (2008).

50. H. He, R.H. Scheicher, R. Pandey, A.R. Rocha, S. Sanvito, A. Grigoriev, R. Ahuja, S.P. Karna, Functionalized nanopore embedded electrodes for rapid DNA sequencing, *J. Phys. Chem. C* 112, 3456–3459 (2008).

51. H. Tanaka, T. Kawai, Visualization of detailed structures within DNA, *Surf. Sci.* 539, L531 (2003).
52. M. Xu, R.G. Endres, Y. Arakawa, The electronic property of DNA bases, *Small* 3, 1539–1543 (2007).
53. E. Shapir, H. Cohen, A. Calzolari, C. Cavazzoni, D.A. Ryndyk, G. Cuniberti, A. Kotlyar, R. Di Felice, D. Porath, Electronic structure of single DNA molecules resolved by transverse scanning tunneling spectroscopy, *Nat. Mater.* 7, 68–74 (2008).
54. C. Lee, X. Wei, J.W. Kysar, J. Hone, Measurement of the elastic properties and intrinsic strength of monolayer graphene, *Science* 321, 385–388 (2008).
55. J.S. Bunch, S.S. Verbridge, J.S. Alden, A.M. Van der Zande, J.M. Parpia, H.G. Craighead, P.L. McEuen, Impermeable atomic membranes from graphene sheets, *Nano Lett.* 8, 2458–2462 (2008).
56. M. Poot, H.S.J. Van der Zant, Nanomechanical properties of few-layer graphene membranes, *Appl. Phys. Lett.* 92, 063111 (2008).
57. L. Tapaszto, G. Dobrik, P. Lambin, L.P. Biro, Tailoring the atomic structure of graphene nanoribbons by scanning tunneling microscope lithography, *Nat. Nanotechnol.* 3, 397–401 (2008).
58. L.C. Venema, J.W.G. Wildoer, H.L.J.T. Tuinstra, C. Dekker, A.G. Rinzler, R.E. Smalley, Length control of individual carbon nanotubes by nanostructuring with a scanning tunneling microscope, *Appl. Phys. Lett.* 71, 2629–2631 (1997).
59. B. Standley, W. Bao, H. Zhang, J. Bruck, C.N. Lau, M. Bockrath, Graphene based atomic scale switches, *Nano Lett.* 8, 3345–3349 (2008).
60. L. Weng, L. Zhang, Y.P. Chen, L.P. Rokhinson, Atomic force microscope local oxidation nanolithography of graphene, *Appl. Phys. Lett.* 93, 093107 (2008).
61. M.D. Fischbein, M. Drndi, Electron beam nanosculpting of suspended graphene sheets, *Appl. Phys. Lett.* 93, 113107 (2008).
62. S.S. Datta, D.R. Strachan, S.M. Khamis, A.T.C. Johnson, Crystallographic etching of few-layer graphene, *Nano Lett.* 8, 1912–1915 (2008).
63. L. Ci, Z. Xu, L. Wang, W. Gao, F. Ding, K. Kelly, B. Yakobson, P. Ajayan, Controlled nanocutting of graphene, *Nano Res.* 1, 116–122 (2008).

64. A. Meller, L. Nivon, D. Branton, Voltage-driven DNA translocations through a nanopore, *Phys. Rev. Lett.* 86, 3435–3438 (2001).
65. A. Storm, C. Storm, J. Chen, H. Zandbergen, J. Joanny, C. Dekker, Fast DNA translocation through a solid-state nanopore, *Nano Lett.* 5, 1193–1197 (2005).
66. J. Chauwin, G. Oster, B.S. Glick, Strong precursor-pore interactions constrain models for mitochondrial protein import, *Biophys. J.* 74, 1732–1743 (1998).
67. S. Ghosal, Electrokinetic flow induced viscous drag on a tethered DNA inside a nanopore, *Phys. Rev. E* 76, 061916 (2007).
68. J. Zhang, B.I. Shklovskii, Effective charge and free energy of DNA inside an ion channel, *Phys. Rev. E* 75, 021906 (2007).
69. T. Hu, B.I. Shklovskii, Theory of DNA translocation through narrow ion channels and nanopores with charged walls, *Phys. Rev. E* 78, 032901 (2008).
70. G. Sigalov, J. Comer, G. Timp, A. Aksimentiev, Detection of DNA sequences using alternating electric field in a nanopore capacitor, *Nano Lett.* 8, 56–63 (2008).
71. J.S. Bunch, A.M. Van der Zande, S.S. Verbridge, I.W. Frank, D.M. Tanenbaum, J.M. Parpia, et al., Electromechanical resonators from graphene sheets, *Science* 315, 490–493 (2007).
72. J. Lagerqvist, M. Zwolak, M. Di Ventra, Comment on "Characterization of the tunneling conductance across DNA bases," *Phys. Rev. E* 76, 013901 (2007).
73. M.E. Gracheva, A. Aksimentiev, J. Leburton, Electrical signatures of single-stranded DNA with single base mutations in a nanopore capacitor, *Nanotechnology* 17, 3160–3165 (2006).
74. M.E. Gracheva, A. Xiong, A. Aksimentiev, K. Schulten, G. Timp, J.P. Leburton, Simulation of the electric response of DNA translocation through a semiconductor nanopore-capacitor, *Nanotechnology* 17, 622–633 (2006).
75. M.J. Allen, V.C. Tung, R.B. Kaner, Honey comb graphene: A review of graphene, *Chem. Rev.* 110, 132 (2010).
76. X. Wang, L.J. Zhi, K. Mullen, Transparent, conductive electrodes for dye sensitized solar cells, *Nano Lett.* 8, 323–327 (2008).
77. G. Eda, G. Fanchini, M. Chhowalla, Large-area ultrathin films of reduced graphene oxide as a transparent and flexible electronic material, *Nat. Nanotechnol.* 3, 270–274 (2008).
78. G. Eda, Y.Y. Lin, S. Miller, C.W. Chen, W.F. Su, M. Chhowalla, Transparent and conducting electrodes for organic electronics from reduced graphene oxide, *Appl. Phys. Lett.* 92, 233305 (2008).

79. M.D. Stoller, S. Park, Y. Zhu, J. An, R.S. Ruoff, Graphene based ultracapacitors, *Nano Lett.* 8, 3498–3502 (2008).
80. R. Kotz, M. Carlen, Principles and applications of electrochemical capacitors, *Electrochim. Acta* 45, 2483–2498 (2000).
81. A. Burke, Ultracapacitors: why, how, and where is the technology? *J. Power Sources,* 91, 37–50 (2000).
82. *Basic Research Needs for Electrical Energy Storage: Report of the Basic Energy Sciences Workshop on Electrical Energy Storage; April 2–4, 2007.* Office of Basic Energy Sciences, Department of Energy, Washington, DC, July 2007.
83. A.G. Pandolfo, A.F. Hollenkamp, Carbon properties and their role in supercapacitors, *J. Power Sources* 157, 11–27 (2006).
84. S. Stankovich, D.A. Dikin, G.H.B. Dommett, K.M. Kohlhaas, E.J. Zimney, E.A. Stach, R.D. Piner, S.T. Nguyen, R.S. Ruoff, Graphene-based composite materials, *Nature* 442, 282–286 (2006).
85. A.K. Geim, P. Kim, Carbon wonderland, *Sci. Am.* 298, 90–97 (2008).
86. R. Ruoff, Calling all chemists, *Nat. Nanotechnol.* 3, 10–11 (2008).
87. S. Stankovich, R.D. Piner, X.Q. Chen, N.Q. Wu, S.T. Nguyen, R.S. Ruoff, Stable aqueous dispersions of graphitic nanoplatelets via the reduction of exfoliated graphite oxide in the presence of poly(sodium 4-styrenesulfonate), *J. Mater. Chem.* 16, 155–158 (2006).
88. D. Li, M. Muller, S. Gilje, R. Kaner, G. Wallace, Processable aqueous dispersions of graphene nanosheets, *Nat. Nanotechnol.* 3, 101–105 (2008).
89. D.A. Dikin, Preparation and characterization of graphene oxide paper, *Nature* 448, 457–460 (2007).
90. S. Park, K.S. Lee, G. Bozoklu, W. Cai, S.T. Nguyen, R.S. Ruoff, Graphene oxide papers modified by divalent ions—Enhancing mechanical properties via chemical cross-linking, *ACS Nano* 2, 572–578 (2008).
91. S. Stankovich, D.A. Dikin, G.H.B. Dommett, K.M. Kohlhaas, E.J. Zimney, E.A. Stach, et al., Graphene-based composite materials, *Nature* 442, 282–286 (2006).
92. S. Watcharotone, D.A. Dikin, S. Stankovich, R. Piner, I. Jung, G.H.B. Dommett, et al., Graphene-silica composite thin films as transparent conductors, *Nano. Lett.* 7, 1888–1892 (2007).

93. V. Khomenko, E. Frackowiak, F. Beguin, Determination of the specific capacitance of conducting polymer/nanotubes composite electrodes using different cell configurations, *Electrochim. Acta* 50, 2499–2506 (2005).

94. G. Lota, T.A. Centeno, E. Frackowiak, F. Stoeckli, Improvement of the structural and chemical properties of a commercial activated carbon for its application in electrochemical capacitors, *Electrochim. Acta* 53, 2210–2216 (2008).

95. S. Stankovich, R.D. Piner, S.T. Nguyen, R.S. Ruoff, Synthesis and exfoliation of isocyanate-treated graphene oxide nanoplatelets, *Carbon* 44, 3342–3347 (2006).

96. M. Winter, R.J. Brodd, What are batteries, fuel cells, and supercapacitors? *Chem. Rev.* 104, 4245–4269 (2004).

97. X. Yu, S. Ye, Recent advances in activity and drability enhancement of Pt/C catalytic cathode in PEMFC Part II: Degradation mechanism and durability enhancement of carbon supported platinum catalyst, *J. Power Sources* 172, 145–154 (2007).

98. K. Gong, F. Du, Z. Xia, M. Dustock, L. Dai, Nitrogen-doped carbon nanotube arrays with high electrocatalytic activity for oxygen reduction, *Science* 323, 760–764 (2009).

99. Smithsonian Institution, Alkali Fuel Cell History, 2001. http://americanhistory.si.edu/fuelcells/alk/alk3.htm.

100. J. Zhang, K. Sasaki, E. Sutter, R.R. Adzic, Stabilization of platinum oxygen-reduction electrocatalysts using gold clusters, *Science* 315, 220–222 (2007).

101. J. Yang, D.J. Liu, N.N. Kariuki, L.X. Chen, Aligned carbon nanotubes with built-in FeN_4 active sites for electrocatalytic reduction of oxygen, *Chem. Commun.* 329–331 (2008).

102. B. Winther-Jensen, O. Winther-Jensen, M. Forsyth, D.R. MacFarlane, High rates of oxygen reduction over a vapor phase-polymerized PEDOT electrode, *Science* 321, 671–674 (2008).

103. J.P. Collman, N.K. Devaraj, R.A. Decréau, Y. Yang, Y.L. Yan, W. Ebina, T.A. Eberspacher, C.E.D.A. Chidsey, Cytochrome c oxidase model catalyzes oxygen to water reduction under rate-limiting electron flux, *Science* 315, 1565–1568 (2007).

104. L. Qu, Y. Liu, J.B. Baek, L. Dai, Nitrogen-doped graphene as efficient metal-free electrocatalysts for oxygen reduction in fuel cells, *ACS Nano* 4, 1321–1326 (2010).

105. K.S. Kim, Y. Zhao, H. Jang, S.Y. Lee, J.M. Kim, K.S. Kim, et al., Large-scale pattern growth of graphene films for stretchable transparent electrodes, *Nature* 457, 706–710 (2009).

106. A. Reina, X. Jia, J. Ho, D. Nezich, H. Son, V. Bulovic, et al., Large area few-layer graphene films on arbitrary substrates by chemical vapor deposition, *Nano Lett.* 9, 30–35 (2009).

107. A.C. Ferrari, J.C. Meyer, V. Scardaci, C. Casiraghi, M. Lazzeri, F. Mauri, et al., Raman spectrum of graphene and graphene layers, *Phys. Rev. Lett.* 97, 187401 (2006).

108. Y. Baskin, L. Meyer, Lattice constants of graphite at low temperatures, *Phys. Rev.* 100, 544 (1955).

109. Y. Xie, P.M.A. Sherwood, Ultrahigh purity graphite electrode by core level and valence band XPS, *Surf. Sci. Spectra* 1, 367–372 (1993).

110. Q. Chen, L. Dai, M. Gao, S. Huang, A. Mau, Plasma activation of carbon nanotubes for chemical modification, *J. Phys. Chem. B* 105, 618–622 (2001).

111. X. Wang, X. Li, L. Zhang, Y. Yoon, P.K. Weber, H. Wang, et al., N-Doping of graphene through electrothermal reactions with ammonia, *Science* 324, 768–771 (2009).

112. D. Wei, Y. Liu, Y. Wang, H. Zhang, L. Huang, G. Yu, Synthesis of N-doped graphene by chemical vapor deposition and its electrical properties, *Nano Lett.* 9, 1752–1758 (2009).

113. P.G. Collins, K. Bradley, M. Ishigami, A. Zettl, Extreme oxygen sensitivity of electronic properties of carbon nanotubes, *Science* 287, 1801–1804 (2000).

114. Z. Yang, Y. Xia, R. Mokaya, Aligned N-doped carbon nanotube bundles prepared via CVD using zeolite substrates, *Chem. Mater.* 17, 4502–4508 (2005).

115. S.Y. Kim, J. Lee, C.W. Na, J. Park, K. Seo, B. Kim, N-Doped double-walled carbon nanotubes synthesized by chemical vapor deposition, *Chem. Phys. Lett.* 413, 300–305 (2005).

116. S. Maldonado, K.J. Stevenson, Direct preparation of carbon nanofiber electrodes via pyrolysis of iron(II) phthalocyanine: Electrocatalytic aspects for oxygen reduction, *J. Phys. Chem. B* 108, 11375–11383 (2004).

117. E.H. Yu, K. Scott, R.W. Reeve, Electrochemical reduction of oxygen on carbon supported Pt and Pt/Ru fuel cell electrodes in alkaline solutions, *Fuel Cells* 3, 169–176 (2003).

118. K.S. Novoselov, D. Jiang, F. Schedin, T.J. Booth, V.V. Khotkevich, S.V. Morozov, A.K. Geim, Two dimensional atomic crystals, *Proc. Natl. Acad. Sci. USA* 102, 10451–10453 (2005).

119. J.C. Meyer, A.K. Geim, M.I. Katsnelson, K.S. Novoselov, T.J. Booth, S. Roth, The structure of suspended graphene sheets, *Nature* 446, 60–63 (2007).

Chapter 6

Graphene-Based Materials for Photonic and Optoelectronic Applications

6.1 Introduction

Decades before the discovery of graphene, it was theoretically predicted that graphene will possess extremely high carrier mobility and an ambipolar field effect [1,2]. This prediction geared up the early experiments that wired mechanically exfoliated flakes by e-beam lithography [3,4]. After several experiments with these graphene samples, researchers firmly believe that graphene will be a wonder material for the next generation of semiconductor devices. The extraordinary electronic properties of graphene can be attributed to the extreme quality of its two-dimensional (2D) crystal lattice [5–7]. The extreme quality here corresponds to an unusually low defect density, which typically serve as the scattering centers that inhibit charge transport.

Kim et al. reported a carrier mobility in excess of 200,000 cm²/(Vs) for a single layer of mechanically exfoliated graphene (see Figure 6.1) [8]. Furthermore, in their experiments, they specifically minimized the substrate-induced scattering by etching under the channel to produce graphene completely suspended between gold contacts. At such high carrier mobility, charge transport is essentially ballistic on the micrometer scale at room temperature. This has major implications for the semiconductor industry because it enables, in principle, fabrication of all-ballistic devices even at today's integrated circuit (IC) channel lengths (down to 45 nm).

The next important point about charge transport in graphene is its ambipolarity. Under field-effect configuration, this implies that carriers can be tuned continuously between holes and electrons by supplying the required gate bias. This can be easily visualized given the unique band structure of graphene (see Figure 6.2) [9]. Under negative gate bias, the Fermi level drops below the Dirac point, introducing a significant

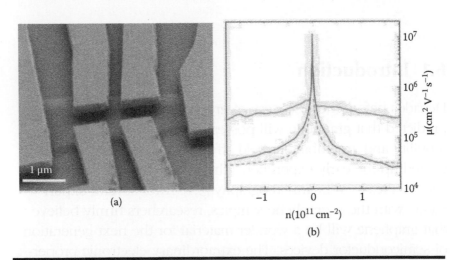

(a)

(b)

Figure 6.1 **Suspended graphene shows extremely high mobility due to the minimization of substrate-induced scattering. (a) Scanning electron microscopic (SEM) image of a suspended sheet after etching. (b) Field-effect measurements indicated mobility greater than 200,000 cm^2/Vs. (Reproduced with permission from K.I. Bolotin, K.J. Sikes, Z. Jiang, M. Klima, G. Fudenberg, J. Hone, et al., *Solid State Commun.* 146, 351–355, 2008. Copyright 2008 Elsevier.)**

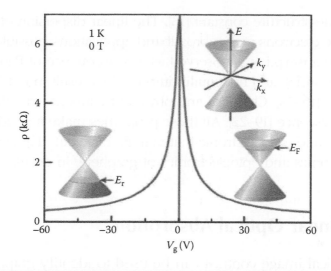

Figure 6.2 Schematic diagram showing the band structure and resulting ambipolar field effect in graphene. Conduction and valence bands meet at the Dirac point without an external field. Under gate bias, the Fermi level moved above or below the Dirac point to introduce a significant number of free carriers. (Reproduced with permission from A.K. Geim, K.S. Novoselov, *Nat. Mater.* 6, 183–191, 2007. Copyright 2007 Nature Publishing Group.)

population of holes into the valence band. Under positive gate bias, the Fermi level rises above the Dirac point, promoting a significant population of electrons into the conduction band. Besides motivating academic interest, access to a truly ambipolar semiconductor enables a number of novel device structures. These are fundamentally different from silicon-based logic because doping levels can be dynamically controlled entirely by gating. By applying local gate biases for a moment to different parts of the same flake, it can form junctions or even more complicated logic. At the same time, rearranging the biases can then completely redefine the device without making any physical changes to the channel of the material.

Graphene exhibits remarkable optical properties, and it can be optically visualized, despite being only a single atom thick [10–12]. Its transmittance T can be expressed in terms

of the fine-structure constant [13]. The linear dispersion of
the Dirac electrons makes broadband applications possible.
Saturable absorption is observed as a consequence of Pauli
blocking [14,15], and nonequilibrium carriers result in hot luminescence [16–18]. Chemical and physical treatments can also lead
to luminescence [19–22]. All these properties make it an ideal
photonic and optoelectronic material. Ferrari et al. [11] reviewed
the photonics and optoelectronics of graphene in detail.

6.2 Linear Optical Absorption

The optical image contrast can be used to identify graphene
on top of an Si/SiO_2 substrate (Figure 6.3a) [12]. This scales
with the number of layers and is the result of interference,
with SiO_2 acting as a spacer. The contrast can be maximized
by adjusting the spacer thickness or the light wavelength
[10,12]. The transmittance of a freestanding single-layer graphene (SLG) can be derived by applying the Fresnel equations
in the thin-film limit for a material with a fixed universal
optical conductance, $G_0 = e2/(4\hbar) \approx 6.08 \times 10^{-5} \ \Omega^{-1}$, to give

$$T = (1 + 0.5 \ \pi\alpha) - 2 \approx 1 - \pi\alpha \approx 97.7\% \tag{6.1}$$

where $\alpha = e^2/(4\pi\varepsilon 0\hbar c) = G_0/(\pi\varepsilon_0 c) \approx 1/137$ is the fine-structure
constant [13]. Graphene only reflects less than 0.1% of the
incident light in the visible region [11,13], rising to about 2%
for 10 layers [12]. Thus, Ferrari et al. [11] assumed that the
optical absorption of graphene layers were proportional to the
number of layers each absorbing $A \approx 1 - T \approx \pi\alpha \approx 2.3\%$ over
the visible spectrum (Figure 6.3b). In a few-layer graphene
(FLG) sample, each sheet can be seen as a 2D electron gas,
with little perturbation from the adjacent layers, making it
optically equivalent to a superposition of almost noninteract-
ing SLG [12]. The absorption spectrum of SLG is quite flat from
300 to 2500 nm with a peak in the ultraviolet region (~270 nm)

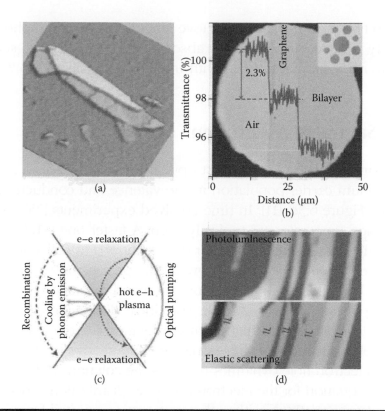

Figure 6.3 The optical properties of graphene: (a) Elastic light-scattering (Rayleigh) image of a graphite flake with varying number of graphene layers. (Reproduced with permission from C. Casiraghi, A. Hartschuh, E. Lidorikis, H. Qian, H. Harutyunyan, T. Gokus, et al., *Nano Lett.* 7, 2711–2717, 2007. Copyright 2007 ACS.) (b) Transmittance for an increasing number of layers. Inset, sample design for the experiment of [13], showing a thick metal support structure with several apertures, on top of which graphene flakes are placed. (Reproduced with permission from Y. Zhang, J.W. Tan, H.L. Stormer, P. Kim, *Nature* 438, 201–204, 2005. Copyright 2008 AAAs.) (c) Schematic of photoexcited electron kinetics in graphene, with possible relaxation mechanisms for the nonequilibrium electron population. (Reproduced with permission from Z. Sun, T. Hasan, F. Torrisi, D. Popa, G. Privitera, F. Wang, F. Bonaccorso, D.M. Basko, A.C. Ferrari, *ACS Nano* 4, 803–810, 2010. Copyright 2010 ACS.) (d) Photoluminescence (top) and elastic scattering (bottom) images of an oxygen-treated flake. (Reproduced with permission from T. Gokus, R.R. Nair, A. Bonetti, M. Bohmler, A. Lombardo, K.S. Novoselov, et al., *ACS Nano* 3, 3963–3968, 2009. Copyright 2009 ACS.) 1L indicates single-layer graphene.

due to the exciton-shifted vanHove singularity in the graphene density of states. In FLG, other absorption features can be seen at lower energies, associated with interband transitions [23,24].

6.3 Saturable Absorption

Interband excitation by ultrafast optical pulses produces a non-equilibrium carrier population in the valence and conduction bands (Figure 6.3c) [11]. In time-resolved experiments [25], two relaxation timescales are typically seen: A faster one (~100 fs) is usually associated with carrier-carrier intraband collisions and phonon emission, and a slower one, on a picosecond timescale, corresponds to electron interband relaxation and cooling of hot phonons [26–28]. The linear dispersion of the Dirac electrons implies that for any excitation there will always be an electron-hole pair in resonance. Quantitative treatment of the electron-hole dynamics requires the solution of the kinetic equation for the electron and hole distribution functions, $f_e(p)$ and $f_h(p)$, with p the momentum counted from the Dirac point [15]. If the relaxation times are shorter than the pulse duration, then during the pulse the electrons reach a stationary state, and collisions put electrons and holes into thermal equilibrium at an effective temperature [15]. The populations determine electron and hole densities, total energy density, and a reduction of photon absorption per layer, due to Pauli blocking, by a factor of $\Delta A/A = [1 - f_e(p)][1 - f_h(p)] - 1$. Assuming efficient carrier-carrier relaxation (both intraband and interband) and efficient cooling of the graphene phonons, the main drawback is energy transfer from electrons to phonons [15]. For linear dispersions near the Dirac point, pair-carrier collisions cannot lead to interband relaxation, thereby conserving the total number of electrons and holes separately [15,29]. Interband relaxation by phonon emission can occur only if the electron and hole energies are close to the Dirac point (within the phonon energy). Radiative recombination of the hot electron-hole population has also been suggested [16–18]. For graphite flakes, the dispersion is quadratic, and pair-carrier collisions can lead to interband relaxation [11]. In principle, decoupled SLG can provide the highest saturable absorption for a given amount of material [15].

6.4 Luminescence

Graphene could be made luminescent by inducing a band gap, following two main routes. One is by cutting it into ribbons and quantum dots; the other is by chemical or physical treatments to reduce the connectivity of the π-electron network. Even though graphene nanoribbons have been produced with varying band gaps [30], as yet no photoluminescence has been reported from them. However, bulk graphene oxide dispersions and solids do show broad photoluminescence [20–22,31]. Individual graphene flakes can be made brightly luminescent by mild oxygen plasma treatment [19]. The resulting photoluminescence is uniform across large areas, as shown in Figure 6.3d, in which a photoluminescence map and the corresponding elastic scattering image are compared. It is possible to make hybrid structures by etching just the top layer, leaving underlying layers intact [19]. This combination of photoluminescent and conductive layers could be used in sandwich light-emitting diodes (LEDs). Luminescent graphene-based materials can now be routinely produced that cover the infrared, visible, and blue spectral ranges [19–22,31]. Although some groups have ascribed photoluminescence in graphene oxide to band gap emission from electron-confined sp^2 islands [20–22], this is more likely to arise from oxygen-related defect states [19]. Whatever the origin, fluorescent organic compounds are of importance to the development of low-cost optoelectronic devices [32]. Blue photoluminescence from aromatic or olefinic molecules is particularly important for display and lighting applications [33].

Luminescent quantum dots are widely used for biolabeling and bioimaging. However, their toxicity and potential environmental hazard limit widespread use and *in vivo* applications. Fluorescent biocompatible carbon-based nanomaterials may be a more suitable alternative. Fluorescent species in the infrared and near-infrared range are useful for biological applications because cells and tissues show little autofluorescence in this region [34]. Sun et al. exploited photoluminescent graphene oxide for live cell imaging in the near-infrared range with little background [21]. Wang et al. have reported a gate-controlled, tunable gap up to 250 meV in bilayer graphene [23]. This may make new photonic devices possible for far-infrared light generation, amplification, and detection.

Broadband nonlinear photoluminescence is also possible following nonequilibrium excitation of untreated graphene layers (Figure 6.3c), as recently reported by several groups [16–18]. Emission occurs throughout the visible spectrum, for energies both higher and lower than the exciting one, in contrast with conventional photoluminescence processes [16–18]. This broadband nonlinear photoluminescence was the consequence of radiative recombination of a distribution of hot electrons and holes, generated by rapid scattering between photoexcited carriers after the optical excitation [16–18], their temperature being determined by interactions with strongly coupled optical phonons [17]. It scales with the number of layers and can be used as a quantitative imaging tool, as well as to reveal the dynamics of the hot electron-hole plasma [16–18] (Figure 6.3c). According to Ferrari et al. [11], for oxygen-induced luminescence, further work is necessary to fully explain this hot luminescence. More recently, electroluminescence was also reported in pristine graphene [36]. However, the power conversion efficiency is lower than it is for carbon nanotubes (CNTs); this could lead to new light-emitting devices based entirely on graphene [11].

6.5 Transparent Conductors

Optoelectronic devices such as displays, touch screens, LEDs, and solar cells require materials with low sheet resistance R_s and high transparency. In a thin film, $R_s = \rho/t$, where t is the film thickness and $\rho = 1/\sigma$ is the resistivity, σ being the direct current (DC) conductivity. For a rectangle of length L and width W, the resistance R is

$$R = \rho/t \times L/W = R_s \times L/W \qquad (6.2)$$

The term L/W can be seen as the number of squares of side W that can be superimposed on the resistor without overlapping. Thus, even if R_s has units of ohms (as R does), it is historically quoted in "ohms per square" ($\Omega/\square$). Current transparent conductors are based on the semiconductors [37] doped indium oxide (In_2O_3) [38], zinc oxide (ZnO) [39], or tin oxide (SnO_2) [37], as well as ternary compounds based on their combinations [37,39,40]. The dominant material is indium tin oxide (ITO), a doped n-type semiconductor composed of about 90% In_2O_3 and about 10% SnO_2 [37].

The electrical and optical properties of ITO are strongly affected by impurities [37]. Tin atoms function as n-type donors [37]. ITO has strong absorption above 4 eV due to interband transitions, with other features at lower energy related to scattering of free electrons by tin atoms or grain boundaries [37]. ITO is commercially available with $T \approx 80\%$ and R_s as low as 10 $\Omega/\square$ on glass72 and about 60–300 $\Omega/\square$ on polyethylene terephthalate [40]. Note that T is typically quoted at 550 nm as this is where the spectral response of the human eye is highest [37]. ITO suffers severe limitations: an ever-increasing cost due to indium scarcity [37], processing requirements, difficulties in patterning [37,40], and sensitivity to both acidic and basic environments. Moreover, it is brittle and can easily wear out or crack when used in applications involving bending, such as touch screens and flexible displays [41].

This means that new transparent conductor materials are needed with improved performance. Metal grids [42], metallic nanowires [43], or other metal oxides [40] have been explored as alternatives. Nanotubes and graphene also show great promise. In particular, graphene films have a higher T over a wider wavelength range than single-wall carbon nanotube (SWNT) films [44–46], thin metallic films [42,43], and ITO [37,39] and are widely preferred in transparent conductors [11].

6.6 Photovoltaic Devices

A photovoltaic cell converts light to electricity [47]. The energy conversion efficiency is given by $\eta = P_{max}/P_{inc}$, where $P_{max} = V_{OC} \times I_{SC} \times FF$ and P_{inc} is the incident power. Here, I_{SC} is the maximum short-circuit current, V_{OC} is the maximum open-circuit voltage, and FF is the fill factor, defined as $FF = (V_{max} \times I_{max})/(V_{OC} \times I_{SC})$, where V_{max} and I_{max} are the maximum voltage and current, respectively [11]. The fraction of absorbed photons converted to current defines the internal photocurrent efficiency. Photovoltaic technology is dominated by silicon cells [47], with η up to about 25% [48]. Organic photovoltaic cells rely on polymers for light absorption and charge transport [49]. They can be manufactured economically compared with silicon cells, for example, by a roll-to-roll process [50], even though they have lower η. An organic photovoltaic cell consists of a transparent conductor, a photoactive layer, and the electrode [49].

Dye-sensitized solar cells use a liquid electrolyte as a charge-transport medium [51]. This type of solar cell consists of a high-porosity nanocrystalline photoanode, comprising TiO_2 and dye molecules, both deposited on a transparent conductor [51]. When illuminated, the dye molecules capture the incident photon, generating electron-hole pairs. The electrons are injected into the conduction band of the TiO_2 and are then transported to the counterelectrode [51,52]. Dye molecules are regenerated by capturing electrons from a liquid electrolyte.

At present, ITO is the most common material for use both as a photoanode and cathode, the latter with a platinum coating. Graphene can fulfill multiple functions in photovoltaic devices: as the transparent conductor window, photoactive material, channel for charge transport, and catalyst. Graphene-based transparent conductive films (GOTCFs) can be used as window electrodes in inorganic (Figure 6.4a), organic (Figure 6.4b), and dye-sensitized (Figure 6.4c) solar cell devices. Wang et al. used GTCFs produced by chemical synthesis, reporting $\eta \approx 0.3\%$ [53].

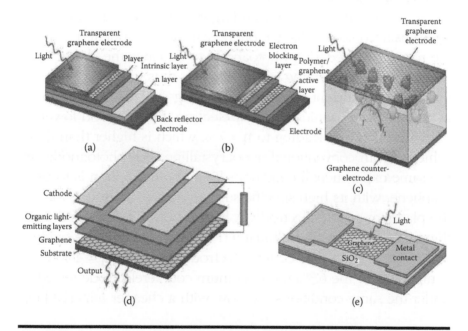

Figure 6.4 Graphene-based optoelectronics. Schematics of (a) inorganic, (b) organic, and (c) dye-sensitized solar cells. (I^- and I^{3-} are iodide and triiodide, respectively). The I^- and I^{3-} ions transfer electrons to the oxidized dye molecules, thus completing the internal electrochemical circuit between the photoanode and the counterelectrode. Schematics of an organic (d) light-emitting diode (LED) and (e) a photodetector. The cylinder in Figure 6.4d represents an applied voltage. (Reproduced with permission from F. Bonaccorso, Z. Sun, T. Hasan, A.C. Ferrari, *Nat. Photonics* 4, 611–622, 2010.)

A higher η of about 0.4% was achieved using reduced graphene oxide, with R_s = 1.6 kΩ/□, instead of 5 kΩ/□, despite a lower T (55% instead of 80%) [53]. De Arco et al. achieved better performance ($\eta \approx 1.2\%$) using chemical vapor deposition (CVD) graphene as the transparent conductor, with R_s = 230 Ω/□ and T = 72% [54]. Further optimization is certainly possible, considering the performance of the best GTCF so far [55]. Graphene oxide dispersions were also used in bulk heterojunction photovoltaic devices, as electron-acceptors with poly(3-hexylthiophene) and poly (3-octylthiophene) as donors, achieving $\eta \approx 1.4\%$ [56]. Yong et al. claimed that $\eta > 12\%$ should be possible with graphene as photoactive material [57].

Graphene can cover an even larger number of functions in dye-sensitized solar cells. Graphene can be incorporated into the nanostructured TiO_2 photoanode to enhance the charge transport rate, preventing recombination, thus improving the internal photocurrent efficiency [58]. Yang et al. used graphene as a TiO_2 bridge, achieving faster electron transport and lower recombination and leading to $\eta \approx 7\%$, which is higher than they achieved with conventional nanocrystalline TiO_2 photoanodes in the same experimental conditions [58]. Another option is to use graphene, with its high specific surface area, to substitute for the platinum counterelectrode. A hybrid poly(3,4-ethylenedioxy thiophene):poly(styrenesulfonate) (PEDOT:PSS)/graphene oxide composite was used as counterelectrode, to obtain η = 4.5%, comparable to the 6.3% for a platinum counterelectrode tested under the same conditions but now with a cheaper material [59].

6.7 Light-Emitting Devices

Organic light-emitting diodes (OLEDs) have an electroluminescent layer between two charge-injecting electrodes, at least one of which is transparent [60]. In these diodes, holes are injected into the highest occupied molecular orbital (HOMO) of the polymer from the anode, and electrons are injected into the lowest unoccupied molecular orbital (LUMO) from the cathode. For an efficient injection, the anode and cathode work functions should match the HOMO and LUMO of the light-emitting polymer [60]. Due to their high image quality, low power consumption, and ultrathin device structure, OLEDs find applications in ultrathin televisions and other display screens, such as computer monitors, digital cameras, and mobile phones. Often, ITO, with its work function about 4.4 to 4.5 eV, is employed as the transparent conductive film. Besides cost issues, ITO is brittle and limited as a flexible substrate [40]. In addition, indium tends to diffuse into the active OLED layers, which reduces device performance over time [37]. Thus, there is a need for alternative transparent conductive films (TCFs) with optical and electrical performance similar to ITO but without its drawbacks.

Graphene has a work function of 4.5 eV, similar to ITO. This, combined with its promise as a flexible and cheap TCF, makes it an ideal candidate for an OLED anode (Figure 6.4d) and further eliminating the issues related to indium diffusion. GTCF anodes enable an outcoupling efficiency comparable to ITO [60]. Considering that the R_s and T of Wu et al. [60], which were 800 $\Omega/\square$ and 82% at 550 nm [60,61], respectively, it is reasonable to expect that further optimization will improve performance.

Matyba et al. [62] used a graphene-oxide based transparent conductive films (GOTCFs) in a light-emitting electrochemical cell. Similar to an OLED, this is a device in which the light-emitting polymer is blended with an electrolyte [61]. The mobile ions in the electrolyte rearrange when a potential is applied between the electrodes, forming layers with high charge density at each electrode interface, which allows efficient and balanced injection of electrons and holes, regardless of the work function of the electrodes [61]..

Usually, the cells have at least one metal electrode. Electrochemical side reactions, involving the electrode materials, can cause problems in terms of operational lifetime and efficiency [62]. This also hinders the development of flexible devices. Thus, graphene is the ideal material to overcome these problems. Matyba et al. [62] demonstrated a light-emitting electrochemical cell based solely on dispersion processable carbon-based materials and opened the door toward totally organic, low-voltage, inexpensive, and efficient LEDs.

6.8 Photodetectors

Photodetectors measure photon flux or optical power by converting the absorbed photon energy into electrical current [11]. They are widely used in a range of common devices [63], such as remote controls, televisions, and DVD players. The internal photoeffect has been exploited to a greater extent, in which the absorption of photons results in carriers excited from the valence to the conduction band, outputting an electric current. The spectral bandwidth is typically limited by the material's absorption [63]. For example, photodetectors based on iv and iii–v semiconductors suffer from the long-wavelength limit, as these become transparent when the incident energy is smaller than the band gap [63]. Graphene absorbs from the ultraviolet-to-terahertz range [64,65]. As a result, graphene-based photodetectors (GPDs; see Figure 6.4e) could work over a much broader wavelength range. The response time is governed by the carrier mobility [63]. Graphene has huge mobilities, so GPDs can be ultrafast.

The photoelectrical response of graphene has been widely investigated both experimentally and theoretically [66–70]. Much broader spectral detection is expected because of the graphene ultrawide band absorption. Xia et al. demonstrated a GPD with a photoresponse of up to 40 GHz [69].

The operating bandwidth of GPDs is mainly limited by their time constant resulting from the device resistance R and capacitance C. Xia et al. reported an RC-limited bandwidth of about 640 GHz [69], which is comparable to traditional photodetectors

[71]. However, the maximum possible operating bandwidth of photodetectors is typically restricted by their transit time, the finite duration of the photogenerated current [63]. The transit-time-limited bandwidth of a GPD could be well over 1500 GHz [69], surpassing state-of-the-art photodetectors. Although an external electric field can produce efficient photocurrent generation with an electron-hole separation efficiency of over 30% [67], zero source-drain bias and dark current operations could be achieved using the internal electric field formed near the metal electrode-graphene interfaces [69,70]. However, the small effective area of the internal electric field could decrease the detection efficiency [69,70] as most of the generated electron-hole pairs would be out of the electric field, thereby recombining rather than being separated. The internal photocurrent efficiencies of about 15–30% [67,68] and external responsivity (generated electric current for a given input optical power) of about 6.1 mA/watt so far reported [70] for GPDs are relatively low compared with current photodetectors [63]. This is mainly due to limited optical absorption when only one SLG is used, short photocarrier lifetimes, and small effective photodetection areas (~200 nm) [71]. The photothermoelectric effect, which exploits the conversion of photon energy into heat and then electric signal [63], may play an important part in photocurrent generation in graphene devices [67,72], and as a result photo-thermoelectric GPDs may be possible.

6.9 Touch Screens

Touch screens are visual outputs that can detect the presence and location of a touch within the display area, permitting physical interaction with what is shown on the display itself [73]. At present, touch panels are used in a wide range of applications, such as cellular phones and digital cameras, because they allow quick, intuitive, and accurate interaction by the user with the display content. Resistive and capacitive touch panels are the most common (Figure 6.5a). A resistive touch panel comprises a conductive substrate, a liquid-crystal device front panel, and a TCF [73]. When pressed, the front-panel film comes into contact with the bottom TCF, and the coordinates of the contact point are calculated on the basis of their resistance values. There are two categories of resistive touch screens: matrix and analogue [73]. The matrix has striped electrodes, whereas the analogue has a nonpatterned transparent conductive electrode with lower production costs.

The TCF requirements for resistive screens are $R_s \approx 500$–$2000\ \Omega/\square$ and $T > 90\%$ at 550 nm [73]. Favorable mechanical

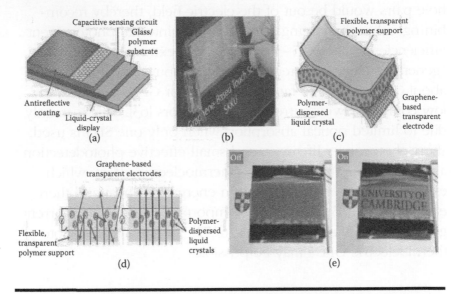

Figure 6.5   Graphene touch screen and smart window. (a) Schematic of a capacitive touch screen. (b) Resistive graphene-based touch screen. (c) Schematic of a PDLC smart window using a GTCF. (d) With no voltage, the liquid-crystal molecules are not aligned, making the window opaque. (e) Graphene/nanotube-based smart window in either an off (left) or on (right) state. [Figure 6.5(a), (c), (d), (e) are all reproduced with permission from F. Bonaccorso, Z. Sun, T. Hasan, A.C. Ferrari, *Nat. Photonics* 4, 611–622, 2010; Figure 6.5(b)—Reproduced with permission from S. Bae, H. Kim, Y. Lee, X. Xu, J.S. Park, Y. Zheng, et al., Roll-to-roll production of 30-inch graphene films for transparent electrodes, *Nature Nanotechnol.* 4, 574–578, 2010.]

properties, including brittleness and wear resistance, high chemical durability, no toxicity, and low production costs, are also important. Cost, brittleness, wear resistance, and chemical durability are the main limitations of ITO [37,40], which cannot withstand the repeated flexing and poking involved with this type of application. Thus, for resistive touch screens there is an effort to find an alternative transparent conductor. GTCFs can satisfy the requirements for resistive touch screens in terms of T and R_s, and exhibit large-area uniformity. Bae et al. [55] reported a graphene-based touch panel display by screen printing a CVD-grown sample (Figure 6.5b). Considering the R_s and T required by analogue resistive screens, GTCFs or GOTCFs produced by liquid phase exfoliation (LPE) are also viable alternatives, further reducing costs.

Capacitive touch screens are emerging as high-end technology, especially since the launch of Apple's iPhone. These consist of an insulator such as glass, coated with ITO [73]. As the human body is also a conductor, touching the surface of the screen results in an electrostatic field distortion, which is measurable as a change in capacitance. Although capacitive touch screens do not work by poking with a pen (making mechanical stresses lower than for resistive screens), the use of GTCFs can still improve performance and reduce costs.

6.10 Flexible Smart Windows and Bistable Displays

Polymer-dispersed liquid-crystal (PDLC) devices were introduced during the 1980s [74, 75]. These consist of thin films of optically transparent polymers with micrometer-size liquid-crystal droplets contained within pores of the polymer. Light passing through the liquid-crystal/polymer is strongly scattered, producing a milky film. If the liquid crystal's ordinary refractive index is close to that of the host polymer, applying an electric field results in a transparent state [41]. In principle, any type of thermotropic liquid crystal may be used in PDLC devices for applications not requiring high switching speeds. More specifically, the ability to switch from translucent to opaque makes them attractive for electrically switchable "smart windows" that can be activated when privacy is required.

Conventionally, ITO on glass is used as the conductive layer to apply the electric field across the PDLC. However, one of the reasons behind the limited market penetration of smart windows is the high cost of ITO. Furthermore, flexibility is hindered when using ITO, reducing potential applications such as PDLC flexible displays [41]. Transparent or colored/tinted smart windows generally require the T to be 60–90% or higher and R_s to be 100–1,000 $\Omega/\square$, depending on production cost, application, and manufacturer. In addition to flexibility, the electrodes need to be as large as the window itself and must have long-term physical and chemical stability, as well as compatibility with the roll-to-roll PDLC production process. Liquid crystals could also be used for next-generation zero-power monochromatic and colored flexible bistable displays, which can retain an image with no power consumption. These are attractive for signs and advertisements or for e-readers and require a transparent flexible conductor for switching the image. The present ITO devices are not ideal for this application owing to the limitations discussed. All these deficiencies of ITO electrodes can be overcome by GTCFs. Figures 6.5c and 6.5d show their working principle, and Figure 6.5e shows a prototype of a flexible smart window with polyethylene terephthalate used as a substrate.

6.11 Saturable Absorbers and Ultrafast Lasers

Materials with nonlinear optical and electro-optical properties are needed in most photonic applications [11]. Laser sources producing nano- to subpicosecond pulses are a key component in the portfolio of leading laser manufacturers [11].

Solid-state lasers have so far been the short-pulse source of choice, being deployed in applications ranging from basic research to materials processing, from eye surgery and printed circuit board manufacturing to metrology and the trimming of electronic components such as resistors and capacitors [11]. Despite wavelength, the majority of ultrafast laser systems use a mode-locking technique; a nonlinear optical element, called a saturable absorber, turns the continuous-wave output into a train of ultrafast optical pulses [76, 77].

The key requirements for nonlinear materials are fast response time, strong nonlinearity, broad wavelength range, low optical loss, high power handling, low power consumption, low cost, and ease of integration into an optical system. Currently, the dominant technology is based on semiconductor saturable absorber mirrors [77]. However, these have a narrow tuning range and require complex fabrication and packaging [14,77]. A simple, cost-effective alternative is to use SWNTs [14,78] in which the diameter controls the gap and thus the operating wavelength. Broadband tunability is possible using SWNTs with a wide diameter distribution [14,78]. However, when operating at a particular wavelength, SWNTs not in resonance are not used and contribute unwanted losses.

As discussed, the linear dispersion of the Dirac electrons in graphene offers an ideal solution: For any excitation, there is always an electron-hole pair in resonance. The ultrafast carrier dynamics [25] combined with large absorption and Pauli blocking make graphene an ideal ultrabroadband, fast saturable absorber. Unlike semiconductor saturable absorber mirrors and SWNTs, there is no need for band gap engineering or chirality/diameter control. So far, graphene-polymer composites [14,15,79–81], CVD-grown films [82,83], functionalized graphene [e.g., graphene oxide bonded with poly (m-phenylenevinylene-co-2,5-dioctoxy-pphenylenevinylene)], and reduced graphene oxide flakes [84,85] have been used for ultrafast lasers. Graphene-polymer composites are scalable and, more important, easily integrated into a range of photonic systems [14,15,79].

Another route for graphene integration is by deterministic placement in a predefined position on a substrate of choice, for example, a fiber core or cavity mirrors. Figure 6.6a shows

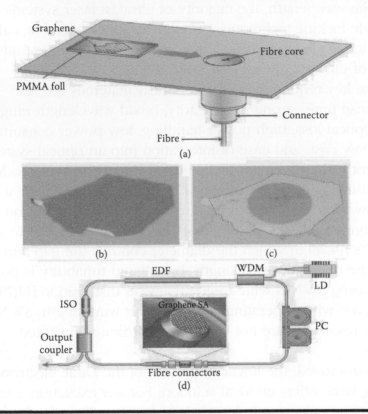

Figure 6.6 Graphene integration in fiber lasers. (a) An optical fiber is mounted onto a holder. Once detached from the original substrate, a polymer/graphene membrane is slid and aligned with the fiber core. (b) Flake originally deposited on SiO₂/Si. (c) The same flake after deterministic placement and dissolution of the polymer layer. (d) Graphene mode-locked ultrafast laser: a graphene saturable absorber (SA) is inserted between two fiber connectors. An erbium-doped fiber (EDF) is the gain medium, pumped by a laser diode (LD) with a wavelength-division multiplexer (WDM). An isolator (ISO) maintains unidirectional operation, and a polarization controller (PC) optimizes mode locking. (Reproduced with permission from F. Bonaccorso, Z. Sun, T. Hasan, A.C. Ferrari, *Nat. Photonics* 4, 611–622, 2010.)

the transfer of such a flake onto an optical fiber core. This is achieved using a water layer between the polymethylmethacrylate (PMMA)/graphene foil and the optical fiber, which enables the PMMA to move. Graphene device integration is finally achieved after precise alignment to the optical fiber core by a micromanipulator (Figure 6.6b) and dissolution of the PMMA layer (Figure 6.6c).

A typical absorption spectrum is shown in Figure 6.7a [14,15,79]. This is featureless apart from the characteristic ultraviolet peak, and the host polymer only contributes a small background for longer wavelengths. Figure 6.7b plots T as a function of average pump power for six wavelengths. Saturable absorption is evident from the T increase with power at all wavelengths. Various strategies have been proposed to integrate graphene saturable absorbers (GSAs) in laser cavities for ultrafast pulse generation. The most common is to sandwich a GSA between two fiber connectors with a fiber adaptor, as shown schematically in Figure 6.6d [14,15,79].

Graphene on a side-polished fiber has also been reported, aimed at high power generation by evanescent field interaction [84–87]. A quartz substrate coated with graphene has been used for free-space solid-state lasers [85, 88]. The most common wavelength of generated ultrafast pulses so far is about 1.5 μm, not because GSAs have any preference for a particular wavelength, but because this is the standard wavelength of optical telecommunications. A solid-state laser mode locked by graphene has been reported at about 1 μm [85]. Figure 6.7c shows a GSA mode-locked laser, made from erbium-doped fiber and tunable from 1526 to 1559 nm, with the tuning range mainly limited by the tunable filter, not the GSA [79]. Figures 6.7d and 6.7e show the pulse from a graphene-oxide-based saturable absorber. The possibility of tuning the GSA properties by functionalization or by using different layers or composite concentrations offers considerable design freedom.

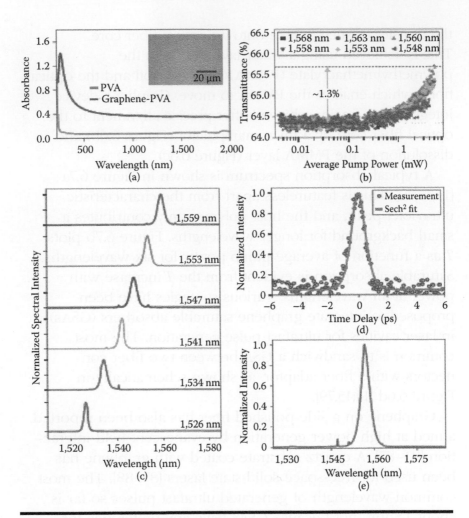

Figure 6.7 **Graphene mode-locked laser performance. (a) Absorption of graphene-PVA (polyvinyl alcohol) composite and reference PVA. Inset: micrograph of the composite. (b) Typical transmittance as a function of pump power at six different wavelengths. (c) Transmittance increases with power. Tunable (>30 nm) fiber laser mode locked by graphene: (d) autocorrelation and (e) spectrum of output pulses of a graphene oxide mode-locked laser, with approximately 743-fs pulse duration. (Figure 6.7 (a), (b) – Requested permission to reproduce image from Z. Sun, T. Hasan, F. Torrisi, D. Popa et al., ACS Nano 4, 803–810, 2010; Fig. 6.7 (c), (d), (e) – Reproduced with permission from F. Bonaccorso, Z. Sun, T. Hasan, A.C. Ferrari, *Nat. Photonics* 4, 611–622, 2010.)**

6.12 Optical Limiters

Optical limiters are devices that have high transmittance for low incident light intensity and low transmittance for high intensity [89]. There is a great interest in these for optical sensors and human eye protection as retinal damage can occur when intensities exceed a certain threshold [89]. Passive optical limiters, which use a nonlinear optical material, have the potential to be simple, compact, and cheap [89]. However, so far no passive optical limiters have been able to protect eyes and other common sensors over the entire visible and near-infrared range [89]. Typical materials include semiconductors (e.g., ZnSe, InSb); organic molecules (e.g., phthalocyanines); liquid crystals; and carbon-based materials (e.g., carbon-black dispersions, CNTs, and fullerenes) [89,90]. Fullerenes and their derivatives [91,92] and CNT dispersions [92] have good optical limiting performance, in particular for nanosecond pulses at 532 and 1064 nm [92]. In graphene-based optical limiters, the absorbed light energy converts into heat, creating bubbles and microplasmas, which results in reduced transmission [90]. Graphene dispersions can be used as wideband optical limiters covering the visible and near-infrared ranges. Broad optical limiting (at 532 and 1064 nm) by LPE graphene was reported for nanosecond pulses [90]. It has also been shown that functionalized graphene dispersions could outperform C60 as an optical limiter [93].

6.13 Optical Frequency Converters

Optical frequency converters are used to expand the wavelength accessibility of lasers (e.g., frequency doubling, parametric amplification and oscillation, and four-wave mixing) [89]. Calculations suggested that nonlinear frequency generation in graphene (harmonics of input light, for example) should be possible for sufficiently high external electric fields

(>100 V cm^{-1}) [94]. Second-harmonic generation from a 150-fs laser at 800 nm has been reported for a graphene film [95]. In addition, four-wave mixing to generate near-infrared wavelength tunable light has been demonstrated using SLG and FLG [96]. Graphene's third-order susceptibility $|\chi^3|$ was measured as about 10^{-7} electrostatic unit (esu) up to one order of magnitude larger than that reported so far for similar measurements on CNTs [96]. However, photon-counting electronics are typically needed to measure the output [95], indicating a low conversion efficiency. Other features of graphene, such as the possibility of tuning the nonlinearity by changing the number of layers [96] and wavelength-independent nonlinear susceptibility, still could be potentially used for various photonic applications.

6.14 Terahertz Devices

Radiation in the 0.3- to 10-THz range (30 μm to 1 mm) is attractive for biomedical imaging, security, remote sensing, and spectroscopy [97]. Much unexplored territory still remains for terahertz technology, mainly owing to a lack of affordable and efficient sources and detectors [97]. The frequency of graphene plasma waves lies in the terahertz range, as well as the gap of graphene nanoribbons, and the bilayer graphene tunable band gap, making graphene appealing for terahertz generation and detection [98]. Various terahertz sources have been suggested based on electrical or optical pumping of graphene devices. Recent experimental observations of terahertz emission and amplification in optically pumped graphene have shown the feasibility of graphene-based terahertz generation [98–100]. Twisted multilayers, retaining the electronic properties of SLG, could also be interesting for such applications. Graphene devices can be used for terahertz detection and frequency conversion. The possibility of tuning the electronic and optical properties by external means (e.g., through electric or magnetic fields or using an optical pump) makes SLG and FLG suitable for infrared and terahertz radiation manipulation. The possible devices include modulators, filters, switches, beam splitters, and polarizers [11].

References

1. P.R. Wallace, The band theory of graphite, *Phys. Rev.* 71, 622–634 (1947).
2. J.C. Slonczewski, P.R. Weiss, Band structure of graphite, *Phys. Rev.* 109, 272–279 (1958).
3. K.S. Novoselov, A.K. Geim, S.V. Morozov, D. Jiang, Y. Zhang, S.V. Dubonos, I.V. Grigorieva, A.A. Firsov, Electric field effect in atomically thin carbon films, *Science* 306, 666–669 (2004).
4. K.S. Novoselov, A.K. Geim, S.V. Morozov, D. Jiang, M.I. Katsnelson, I.V. Grigorieva, S.V. Dubonos, A.A. Firsov, Two-dimensional gas of massless Dirac fermions in graphene, *Nature* 438, 197–200 (2005).
5. D. Jiang, F. Schedin, T.J. Booth, V.V. Khotkevich, S.V. Morozov, A.K. Geim, Two-dimensional atomic crystals, *Proc. Natl. Acad. Sci. USA* 102, 10451–10453 (2005).
6. S.V. Morozov, K.S. Novoselov, M.I. Katsnelson, F. Schedin, D.C. Elias, J.A. Jaszczak, et al., Giant intrinsic carrier mobilities in graphene and its bilayer, *Phys. Rev. Lett.* 100, 016602 (2008).
7. K.S. Novoselov, S.V. Morozov, T.M.G. Mohinddin, L.A. Ponomarenko, D.C. Elias, R. Yang, et al., Electronic properties of graphene, *Phys. Status Solidi B: Basic Solid State Phys.* 244, 4106–4111 (2007).
8. K.I. Bolotin, K.J. Sikes, Z. Jiang, M. Klima, G. Fudenberg, J. Hone, et al., Ultrahigh electron mobility in suspended graphene, *Solid State Commun.* 146, 351–355 (2008).
9. A.K. Geim, K.S. Novoselov, The rise of graphene, *Nat. Mater.* 6, 183–191 (2007).
10. P. Blake, E.W. Hill, A.H.C. Neto, K.S. Novoselov, D. Jiang, R. Yang, et al., Making graphene visible, *Appl. Phys. Lett.* 91, 063124 (2007).
11. F. Bonaccorso, Z. Sun, T. Hasan, A.C. Ferrari, Graphene photonics and optoelectronics, *Nat. Photonics* 4, 611–622 (2010).

12. C. Casiraghi, A. Hartschuh, E. Lidorikis, H. Qian, H. Harutyunyan, T. Gokus, et al., Rayleigh imaging of graphene and graphene layers, *Nano Lett.* 7, 2711–2717 (2007).

13. R.R. Nair, P. Blake, A.N. Grigorenko, K.S. Novoselov, T.J. Booth, T. Stauber, et al., Fine structure constant defines visual transparency of graphene, *Science* 320, 1308 (2008).

14. T. Hasan, Z. Sun, F. Wang, F. Bonaccorso, P.H. Tan, A.G. Rozhi, et al., Nanotube-polymer composites for ultrafast photonics, *Adv. Mater.* 21, 3874–3899 (2009).

15. Z. Sun, T. Hasan, F. Torrisi, D. Popa, G. Privitera, F. Wang, F. Bonaccorso, D.M. Basko, A.C. Ferrari, Graphene mode-locked ultrafast laser, *ACS Nano* 4, 803–810 (2010).

16. R.J. Stoehr, R. Kolesov, J. Pflaum, J. Wrachtrup, Fluorescence of laser created electron-hole plasma in graphene, *Phys. Rev. B* 82, 121408(R) (2010).

17. C.H. Liu, K.F. Mak, J. Shan, T.F. Heinz, Ultrafast photoluminescence from graphene, *Phys. Rev. Lett.* 105, 127404 (2010).

18. W. Liu, S.W. Wu, P.J. Schuck, M. Salmeron, Y.R. Shen, F. Wang, Nonlinear photoluminescence from graphene, Abstract number: BAPS. 2010. MAR. Z22. 11, APS March Meeting, Portland, OR (2010).

19. T. Gokus, R.R. Nair, A. Bonetti, M. Bohmler, A. Lombardo, K.S. Novoselov, et al., Making graphene luminescent by oxygen plasma treatment, *ACS Nano* 3, 3963–3968 (2009).

20. G. Eda, Y.Y. Lin, C. Mattevi, H. Yamaguchi, H.A. Chen, I.S. Chen, et al., Blue photoluminescence from chemically derived graphene oxide, *Adv. Mater.* 22, 505–509 (2009).

21. X. Sun, Z. Liu, K. Welsher, J.T. Robinson, A. Goodwin, S. Zaric, et al., Nano-graphene oxide for cellular imaging and drug delivery, *Nano Res.* 1, 203–212 (2008).

22. Z. Luo, P.M. Vora, E.J. Mele, A.T. Johnson, J.M. Kikkawa, Photoluminescence and band gap modulation in graphene oxide, *Appl. Phys. Lett.* 94, 111909 (2009).

23. Y. Zhang, J.W. Tan, H.L. Stormer, P. Kim, Experimental observation of the quantum Hall effect and Berry's phase in graphene, *Nature* 438, 201–204 (2005).

24. K.F. Mak, J. Shan, T.F. Heinz, Electronic structure of few-layer graphene: Experimental demonstration of strong dependence on stacking sequence, *Phys. Rev. Lett.* 104, 176404 (2009).

25. M. Breusing, C. Ropers, T. Elsaesser, Ultrafast carrier dynamics in graphite, *Phys. Rev. Lett.* 102, 086809 (2009).
26. T. Kampfrath, L. Perfetti, F. Schapper, C. Frischkorn, M. Wolf, Strongly coupled optical phonons in the ultrafast dynamics of the electronic energy and current relaxation in graphite, *Phys. Rev. Lett.* 95, 187403 (2005).
27. M. Lazzeri, S. Piscanec, F. Mauri, A.C. Ferrari, J. Robertson, Electronic transport and hot phonons in carbon nanotubes, *Phys. Rev. Lett.* 95, 236802 (2005).
28. Z. Sun, T. Hasan, F. Torrisi, D. Popa, G. Privitera, F. Wang, et al., Graphene mode-locked ultrafast laser, *ACS Nano* 4, 803–810 (2010).
29. J. Gonzalez, F. Guinea, M.A.H. Vozmediano, Unconventional quasiparticle lifetime in graphite, *Phys. Rev. Lett.* 77, 3589–3592 (1996).
30. M.Y. Han, B. Ozyilmaz, Y. Zhang, P. Kim, Energy band-gap engineering of graphene nanoribbons. *Phys. Rev. Lett.* 98, 206805 (2007).
31. J. Lu, J.X. Yang, J. Wang, A. Lim, S. Wang, P. Loh, One-pot synthesis of fluorescent carbon nanoribbons, nanoparticles, and graphene by the exfoliation of graphite in ionic liquids, *ACS Nano* 3, 2367–2375 (2009).
32. J.R. Sheats, H. Antoniadas, M. Hueschen, W. Leonard, J. Miller, R. Moon, et al., Organic electroluminescent devices, *Science* 273, 884–888 (1996).
33. L.J. Rothberg, A.J. Lovinger, Status of and prospects for organic electroluminescence, *J. Mater. Res.* 11, 3174–3187 (1996).
34. J.V. Frangioni, *In vivo* near-infrared fluorescence imaging, *Curr. Opin. Chem. Biol.* 7, 626–634 (2003).
35. K.Welsher, Z. Liu, D. Daranciang, H. Dai, Selective Probing and Imaging of Cells with SWCNTs as Near-Infra red Fluorescent Molecules, *Nano. Lett* 8, 586–590 (2008).
36. S. Essig, C.W. Marquardt, A. Vijayaraghavan, M. Ganzhorn, S. Dehm, F. Hennrich, et al., Phonon-assisted electroluminescence from metallic carbon nanotubes and graphene, *Nano Lett.* 10, 1589–1594 (2010).
37. I. Hamberg, C.G. Granqvist, Evaporated Sn-doped In_2O_3 films: Basic optical properties and applications to energy-efficient windows, *J. Appl. Phys.* 60, R123–R160 (1986).
38. L. Holland, G. Siddall, The properties of some reactively sputtered metal oxide films, *Vacuum* 3, 375–391 (1953).

39. T. Minami, Transparent conducting oxide semiconductors for transparent electrodes, *Semicond. Sci. Technol.* 20, S35–S44 (2005).

40. C.G. Granqvist, Transparent conductors as solar energy materials: A panoramic review, *Sol. Energy Mater. Sol. Cells* 91, 1529–1598 (2007).

41. C.D. Sheraw, L. Zhou, J.R. Huang, D.J. Gundlach, T.N. Jackson, Organic thin-film transistor-driven polymer dispersed liquid crystal displays on flexible polymeric substrates, *Appl. Phys. Lett.* 80, 1088–1090 (2002).

42. J.Y. Lee, S.T. Connor, Y. Cui, P. Peumans, Solution-processed metal nanowire mesh transparent electrodes, *Nano Lett.* 8, 689–692 (2008).

43. S. De, T.M. Higgins, P.E. Lyons, E.M. Doherty, P.N. Nirmalraj, W.J. Blau, et al., Silver nanowire networks as flexible, transparent, conducting films: Extremely high dc to optical conductivity ratios, *ACS Nano* 3, 1767–1774 (2009).

44. H.-Z. Geng, K.-K. Kim, K.-P. So, Y.S. Lee, Y. Chang, Y.H. Lee, Effect of acid treatment on carbon nanotube-based flexible transparent conducting films, *J. Am. Chem. Soc.* 129, 7758–7759 (2007).

45. Z. Wu, Z. Chen, X. Du, J.M. Logan, J. Sippel, M. Nikolou, et al., Transparent, conductive carbon nanotube films, *Science* 305, 1273–1276 (2004).

46. S. De, J.N. Coleman, Are there fundamental limitations on the sheet resistance and transmittance of thin graphene films, *ACS Nano* 4, 2713–2720 (2010).

47. D.M. Chapin, C.S. Fuller, G.L. Pearson, A new silicon p-n junction photocell for converting solar radiation into electrical power, *J. Appl. Phys.* 25, 676–677 (1954).

48. M.A. Green, K. Emery, K. Bucher, D.L. King, S. Igari, Solar cell efficiency table, *Prog. Photovolt. Res. Appl.* 7, 321–326 (1999).

49. H. Hoppe, N.S. Sariciftci, Organic solar cells: An overview, *MRS Bull.* 19, 1924–1945 (2004).

50. F.C. Krebs, All solution roll-to-roll processed polymer solar cells free from indium-tin-oxide and vacuum coating steps. *Org. Electron.* 10, 761–768 (2009).

51. B. O'Regan, M.A. Gratzel, Low-cost, high-efficiency solar cell based on dye-sensitized colloidal TiO_2 films, *Nature* 353, 737–740 (1991).

52. J. Wu, H.A. Becerril, Z. Bao, Z. Liu, Y. Chen, P. Peumans, Organic solar cells with solution-processed graphene transparent electrodes, *Appl. Phys. Lett.* 92, 263302 (2008).

53. X. Wang, L. Zhi, N. Tsao, Z. Tomovic, J. Li, K. Mullen, Transparent carbon films as electrodes in organic solar cells, *Angew. Chem.* 47, 2990–2992 (2008).

54. L.G. De Arco, Y. Zhang, C.W. Schlenker, K. Ryu, M.E. Thompson, C. Zhou, Continuous, highly flexible and transparent graphene films by chemical vapor deposition for organic photovoltaics, *ACS Nano* 4, 2865–2873 (2010).

55. S. Bae, H. Kim, Y. Lee, X. Xu, J.S. Park, Y. Zheng, et al., Roll-to-roll production of 30-inch graphene films for transparent electrodes, *Nature Nanotechnol.* 4, 574–578 (2010).

56. Z. Liu, Q. Liu, Y. Huang, Y. Ma, S. Yin, X. Zhang, et al., Organic photovoltaic devices based on a novel acceptor material: graphene, *Adv. Mater.* 20, 3924–3930 (2008).

57. V. Yong, J.M. Tour, Theoretical efficiency of nanostructured graphene based photovoltaics, *Small* 6, 313–318 (2009).

58. W. Hong, Y. Xu, G. Lu, C. Li, G. Shi, Transparent graphene/PEDOT-PSS composite films as counter electrodes of dye sensitized solar cells, *Electrochem. Commun.* 10, 1555–1558 (2008).

59. J.H. Burroughes, D.D.C. Bradley, A.R. Brown, R.N. Marks, K. Mackay, R.H. Friend, et al., Light-emitting diodes based on conjugated polymers, *Nature* 347, 539–541 (1990).

60. J. Wu, M. Agarwal, H.A. Becerril, Z. Bao, Z. Liu, Y. Chen, et al., Organic light-emitting diodes on solution-processed graphene transparent electrodes, *ACS Nano* 4, 43–48 (2009).

61. Q. Pei, A.J. Heeger, Operating mechanism of light-emitting electrochemical cells, *Nat. Mater.* 7, 167 (2008).

62. P. Matyba, H. Yamaguchi, G. Eda, M. Chhoalla, L. Edman, N.D. Robinson, Graphene and mobile ions: The key to all-plastic, solution processed light-emitting devices, *ACS Nano* 4, 637–642 (2010).

63. B.E.A. Saleh, M.C. Teich, *Fundamentals of Photonics,* pp. 784–803, Wiley, New York (2007).

64. J.M. Dawlaty, S. Shivaram, J. Strait, P. George, M.V.S. Chandrashekhar, F. Rana, et al., Measurement of the optical absorption spectra of epitaxial graphene from terahertz to visible, *Appl. Phys. Lett.* 93, 131905 (2008).

65. A.R. Wright, J.C. Cao, C. Zhang, Enhanced optical conductivity of bilayer graphene nanoribbons in the terahertz regime, *Phys. Rev. Lett.* 103, 207401 (2009).

66. F.T. Vasko, V. Ryzhii, Photoconductivity of intrinsic graphene, *Phys. Rev. B* 77, 195433 (2008).

67. J. Park, Y.H. Ahn, C. Ruiz-Vargas, Imaging of photocurrent generation and collection in single-layer graphene, *Nano Lett.* 9, 1742–1746 (2009).

68. F. Xia, T. Mueller, R.G. Mojarad, M. Freitag, Y.M. Lin, J. Tsang, et al., Photocurrent imaging and efficient photon detection in a graphene transistor, *Nano Lett.* 9, 1039–1044 (2009).

69. F. Xia, T. Mueller, Y.M. Lin, A. Valdes-Garcia, P. Avouris, Ultrafast graphene photodetector, *Nat. Nanotech.* 4, 839–843 (2009).

70. T. Mueller, F. Xia, P. Avouris, Graphene photodetectors for high-speed optical communications, *Nat. Photon.* 4, 297–301 (2010).

71. Y. Kang, H.D. Liu, M. Morse, M.J. Paniccia, M. Zadka, S. Litski, et al., Monolithic germanium/silicon avalanche photodiodes with 340 GHz gain-bandwidth product, *Nat. Photon.* 3, 59–63 (2009).

72. X.D. Xu, N.M. Gabor, J.S. Alden, A.M. Van der Zande, P.L. McEuen, Photo-thermoelectric effect at a graphene interface junction, *Nano Lett.* 10, 562 (2010).

73. J.A. Pickering, Touch-sensitive screens: The technologies and their applications, *Int. J. Man. Mach. Stud.* 25, 249–269 (1986).

74. H.G. Craighead, J. Cheng, S. Hackwood, New display based on electrically induced index-matching in an inhomogeneous medium, *Appl. Phys. Lett.* 40, 22–24 (1982).

75. J.L. Fergason, Polymer Encapsulated Nematic Liquid Crystals for Display and Light ControlApplications, *SID Symposium Digest* 16, 68–70 (1985).

76. T. Hasan, Z. Sun, F. Wang, F. Bonaccorso, P.H. Tan, A.G. Rozhi, A.C. Ferrari, Nanotube-polymer composites for ultrafast photonics, *Adv. Mater.* 21, 3874–3899 (2009).

77. U. Keller, Recent developments in compact ultrafast lasers, *Nature* 424, 831–838 (2003).

78. F. Wang, A.G. Rozhin, V. Scardaci, Z. Sun, F. Hennrich, I.H. White, W.I. Milne, A.C. Ferrari, Wideband-tunable, nanotube mode-locked, fibre laser, *Nat. Nanotechnol.* 3, 738–742 (2008).

79. Z. Sun, D. Popa, T. Hasan, F. Torrisi, F. Wang, E.J.R. Kelleher, et al., Wideband tunable, graphene-mode locked, ultrafast laser, *Nano Res.* 3, 653 (2010).

80. Q. Bao, H. Zhang, Y. Wang, Z. Ni, Y. Yan, Z.X. Shen, et al., Atomic-layer graphene as a saturable absorber for ultrafast pulsed lasers, *Adv. Funct. Mater.* 19, 3077–3083 (2010).

81. H. Zhang, Q.L. Bao, D.Y. Tang, L.M. Zhao, K. Loh, Large energy soliton erbium-doped fiber laser with a graphene-polymer composite mode locker, *Appl. Phys. Lett.* 95, 141103 (2009).

82. H. Zhang, D.Y. Tang, L.M. Zhao, Q.L. Bao, K.P. Loh, Large energy mode locking of an erbium-doped fiber laser with atomic layer graphene, *Opt. Express* 17, 17630–17635 (2009).

83. H. Zhang, D. Tang, R.J. Knize, L. Zhao, Q. Bao, K.P. Loh, Graphene mode locked, wavelength-tunable, dissipative soliton fiber laser, *Appl. Phys. Lett.* 96, 111112 (2010).

84. Y.W. Song, S.Y. Jang, W.S. Han, M.K. Bae, Graphene mode-lockers for fiber lasers functioned with evanescent field interaction, *Appl. Phys. Lett.* 96, 051122 (2010).

85. W.D. Tan, C.Y. Su, R.J. Knize, G.Q. Xie, L.J. Li, D.Y. Tang, Mode locking of ceramic Nd:yttrium aluminum garnet with graphene as a saturable absorber, *Appl. Phys. Lett.* 96, 031106 (2010).

86. V. Scardaci, Z. Sun, F. Wang, A.G. Rozhin, T. Hasan, F. Hennrich, et al., Carbon nanotube polycarbonate composites for ultrafast lasers, *Adv. Mater.* 20, 4040–4043 (2008).

87. Z. Sun, A.G. Rozhin, F. Wang, T. Hasan, D. Popa, W. O'Neill, et al., A compact, high power, ultrafast laser mode-locked by carbon nanotubes, *Appl. Phys. Lett.* 95, 253102 (2009).

88. Q. Bao, H. Zhang, Z. Ni et al., *Nano Research* 4, 297–207 (2011).

89. M. Bass, G. Li, E.V. Stryland, *Handbook of Optics IV*, McGraw-Hill, New York (2001).

90. J. Wang, Y. Hernandez, M. Lotya, J.N. Coleman, W.J. Blau, Broadband nonlinear optical response of graphene dispersions, *Adv. Mater.* 21, 2430–2435 (2009).

91. L.W. Tutt, A. Kost, Optical limiting performance of C60 and C70 solutions, *Nature* 356, 225–226 (1992).

92. J. Wang, Y. Chen, W.J. Blau, Carbon nanotubes and nanotube composites for nonlinear optical devices, *J. Mater. Chem.* 19, 7425–7443 (2009).

93. Y. Xu, Z. Liu, X. Zhang, Y. Wang, J. Tian, Y. Huang, Y. Ma, X. Zhang, Y. Chen, A graphene hybrid material covalently functionalized with porphyrin: synthesis and optical limiting property, *Adv. Mater.* 21, 1275–1279 (2009).

94. S.A. Mikhailov, Non-linear electromagnetic response of graphene, *Europhys. Lett.* 79, 27002 (2007).

95. J.J. Dean, H.M. Van Driel, Second harmonic generation from graphene and graphitic films, *Appl. Phys. Lett.* 95, 261910 (2009).

96. E. Hendry, P.J. Hale, J.J. Moger, A.K. Savchenko, S.A. Mikhailov, Coherent nonlinear optical response of graphene, *Phys. Rev. Lett.* 105, 097401 (2010).

97. X.C. Zhang, J. Xu, *Introduction to THz Wave Photonics*, Springer, New York (2010).

98. F. Rana, Graphene terahertz plasmon oscillators, *IEEE Trans. Nanotechnol.* 7, 91–99 (2008).

99. D. Sun, C. Divin, J. Rioux, J.E. Sipe, C. Berger, W.A. De Heer, et al., Coherent control of ballistic photocurrents in multilayer epitaxial graphene using quantum interference, *Nano Lett.* 10, 1293–1296 (2010).

100. T. Otsuji, H. Karasawa, T. Komori, T. Watanabe, M. Suemitsu, A. Satou, et al., Observation of amplified stimulated terahertz emission from optically pumped epitaxial graphene heterostructures, PIERS Proceedings, Xi'an, China, March 22–26, 2010.

Index

Printed and bound by CPI Group (UK) Ltd, Croydon, CR0 4YY

18/10/2024

01776262-0001